INNOVATION AND STRATEGY OF ONLINE GAMES

Series on Technology Management*

Series Editor: J. Tidd (Univ. of Sussex, UK) ISSN 0219-9823

Published

Vol. 4 Japanese Cost Management
edited by Y. Monden *(Univ. of Tsukuba, Japan)*

Vol. 5 R&D Strategy on Organisation
Managing Technical Change in Dynamic Contexts
by V. Chiesa *(Univ. degli Studi di Milano, Italy)*

Vol. 6 Social Interaction and Organisational Change
Aston Perspectives on Innovation Networks
edited by O. Jones *(Aston Univ., UK)*, S. Conway *(Aston Univ., UK)* & F. Steward *(Aston Univ., UK)*

Vol. 7 Innovation Management in the Knowledge Economy
edited by B. Dankbaar *(Univ. of Nijmegen, The Netherlands)*

Vol. 8 Digital Innovation
Innovation Processes in Virtual Clusters and Digital Regions
edited by G. Passiante (*Univ. of Lecce, Italy*), V. Elia (*Univ. of Lecce, Italy*) & T. Massari (*Univ. of Lecce, Italy*)

Vol. 9 Service Innovation
Organisational Responses to Technological Opportunities and Market Imperatives
edited by J. Tidd (*Univ. of Sussex, UK*) & F. M. Hull (*Fordham Univ., USA*)

Vol. 10 Open Source
A Multidisciplinary Approach
by M. Muffatto (*University of Padua, Italy*)

Vol. 11 Involving Customers in New Service Development
edited by B. Edvardsson, A. Gustafsson, P. Kristensson, P. Magnusson & J. Matthing (*Karlstad University, Sweden*)

Vol. 12 Project-Based Organization in the Knowledge-Based Society
by M. Kodama (*Nihon University, Japan*)

Vol. 13 Building Innovation Capability in Organizations
An International Cross-Case Perspective
by M. Terziovski (*University of Melbourne, Australia*)

Vol. 14 Innovation and Strategy of Online Games
by Jong H Wi (*Chung-Ang University, South Korea*)

*For the complete list of titles in this series, please write to the Publisher.

SERIES ON TECHNOLOGY MANAGEMENT – VOL. 14

INNOVATION AND STRATEGY OF ONLINE GAMES

Jong H Wi

Chung-Ang University, South Korea

ICP Imperial College Press

Published by

Imperial College Press
57 Shelton Street
Covent Garden
London WC2H 9HE

Distributed by

World Scientific Publishing Co. Pte. Ltd.
5 Toh Tuck Link, Singapore 596224
USA office: 27 Warren Street, Suite 401-402, Hackensack, NJ 07601
UK office: 57 Shelton Street, Covent Garden, London WC2H 9HE

British Library Cataloguing-in-Publication Data
A catalogue record for this book is available from the British Library.

INNOVATION AND STRATEGY OF ONLINE GAMES
Series on Technology Management — Vol. 14

ISBN-13 978-1-84816-356-0
ISBN-10 1-84816-356-8

Printed by FuIsland Offset Printing (S) Pte Ltd, Singapore

To my Mother, my wife Pilsun and two daughters,
Yokyoung and Yomin

Preface

In the late 1990s, when the online game industry first began to expand in Korea, I had some opportunities to talk with a few Japanese console game developers. I explained to them the attributes and business model of the online game, which was a new upcoming trend. Unfortunately, however, their reaction was contrary to what I expected. They didn't understand 'What online game is' and I was surprised by their response.

Even in the past 10 years, the attitudes of console game developers to the online game have not changed much. It seems impossible for orthodox console and PC game developers not only to understand the online game, but also to design and develop online games.

Generally speaking, the console game and the online game seem similar in terms of games. Playing styles are very similar. Controlling characters, background game graphics and sounds, and solving quests are similar elements in both. Nevertheless, why can't console game developers grab the essence of the online game? The reason can be found in this term, 'disruptive innovation'.

Online gaming is a new industry derived from the PC and console game through disruptive innovation. Considering the technical aspect, for instance, offline games like the PC and console game are launched in the form of the game package consisting of just a client program; on the other hand, online games comprise not only a client program but also a server program which can hold a lot of access users concurrently. Roles of game servers consist of a number of users ranging from dozens to millions, allowing many users to play the same game at the same time. In this context, as offline game developers do not have any knowledge about server programming, they are not able to construct a game server system. In view of online game business, offline game developers only have 'half capability'.

Furthermore, one of the critical techniques in managing game servers is security. Once online game servers are hacked, cyber assets accumulated by users can be stolen, or free servers managed by unauthorized subjects can result in a leak of essential game source codes. Problems that offline game developers could never have imagined may happen in the online game industry.

Moreover, product attributes between online games and offline games are fundamentally different. In offline games, game quality itself is the most important barometer. In online games, however, besides game contents, community identity creates a new product attribute added to the online game. Community identity refers to a process whereby users evolve game contents while playing an online game. This process shows that users have taken one step into the game developing process, and it can be interpreted in terms of *Democratizing Innovation* by von Hippel.

Therefore, accepting various requirements from users and adapting them into game contents are very important for online game development. However, for offline game developers, users' penetration into game development process is not easy to accept. In the eyes of offline game developers, users are not companions constructing game contents together, but passive consumers just playing games made by them. There are also huge differences in the business model and the distribution channel between both of them.

Having considered these reasons, it is not difficult to understand why offline game developers cannot understand online gaming to the extent that they sometimes feel hostile towards it. Online gaming became a 'service', not just a game, through disruptive innovation.

Explanations about the online game industry based on the innovation theory suggest sound reasons why the online game was first industrialized in Korea instead of in Japan or the USA, two big powers in the game industry. It also explains how Korea came to acquire the best online game development technologies in the world. Compared to comics, animation and the console game business for which Korea was the original equipment manufacturer (OEM) of Japan or the USA, the online game business was conceptualized in Korea, based on strong information technology (IT) infrastructure. It swept the Asian market and has now expanded to the rest of the world. Moreover, diverse businesses such as the game community, Avatar service and the item-based model have gained the interest of global game and Internet companies.

This book is the first study to survey innovations and the industrial formation process of online game business, and global strategies of major Korean online game companies. I have examined diverse aspects of the online game business over a ten-year period. Beginning from the innovation of online games, this book contains many topics broadly related to online gaming, including the main factors stimulating online game business in Korea, comparisons of users' attributes in various countries, differences between online and offline game users, and virtual business created by online games. Research on user attributes and market investigation were conducted over a five-year period in China, Japan, USA and Southeast Asian countries as well.

Chapter 5 focuses especially on an analysis of relevant government policies. Government policies toward businesses always attract controversy. Far from the general opinion reported by the foreign media, the Korean Internet contents industry has not developed as a whole due to support from government IT industry policies. To a creativity-based business like Internet contents, government support can be a double-edged sword. This book clarifies the actual role of the Korean government at the beginning and at the developmental period of the online gaming business.

I sincerely hope that this book will be helpful to readers who have an interest in innovative and creative businesses like the Internet contents business.

Jong H. Wi

Digital Media City, Seoul, Korea
8 January 2009

Contents

Acknowledgements

I was fortunate to have an opportunity to be involved first-hand in the online gaming industry. I say this because online gaming is basically a knowledge-based industry. In addition, this industry has huge potential to influence various related industries. Above all, most people who are engaged in this industry are young and energetic, with enthusiasm in pursuing the future. Therefore, it was a pleasant and exciting experience to have talks with scholars, businessmen, government officials and the media from Korea and other countries while I did research in this field. Once again, I would like to thank all the people whom I have met since I began the study on the online gaming industry.

First of all, I am deeply grateful to Dr. Joseph Tidd (University of Sussex) and Dr. Jonathan Sapsed (University of Brighton). They gave me a lot of advice and support for publishing this book. Without their help, the book could not have been made available to Western readers. Junjiro Shintaku (University of Tokyo), Chaisung Lim (Kunkuk University) and Akira Baba (University of Tokyo) always gave me good comments and ideas in my research. Kiyoshi Shin (IGDA Japan Coordinator) and Kenji Matsubara (CEO, KOEI), who are my good friends, inspired me through discussions on various subjects related to the game industry. I also want to express my gratitude for their friendship.

As for people from the industry, I wish to thank Chris Whang and Joshua Hong (CEO, K2 Network), II Sung Baik and Yangshin Kim (CEO, JC Entertainment), Tony Park (Vice President, The Nine), and Jake Song (CEO, XLgames). Also, Arts Won, Sol Yi, Insoo

Song (Research Managers of Contents Management Institute) and other CMI researchers who have supported my surveys and research around the world with the utmost commitment. I will not forget their dedication to my work.

Jong H. Wi
8 January 2009

About the Author

Jong H. Wi (Ph.D. in Strategic Management, University of Tokyo) is an Associate Professor at Chung-Ang University and President of Game Contents Research Center at Chung-Ang University, Korea. He is also President of the Contents Management Institute, Vice President of Japanese Online Game Focused Association, advisory member of the Korean government and National Assembly, and consultant to many Korean and foreign game companies and governments. He is an expert and well-known researcher on the Asian online game industry, and has published numerous books, academic papers, white papers and business reports.

He has written several books on the gaming and virtual business industry from the perspective of innovation such as *Industrial Development Strategy of Online Games* (Tsinghua University Press, 2008, in Chinese), *Culture Marketing* (Hankyungsa, 2008, in Korean), *Business Strategy of Second Life* (Joong-Ang Books, 2007, in Korean), *Innovation Strategy of Japanese Firms* (Jipmundang, 2007, in Korean), *The Research of Korean Online Game Business* (Toyokeizaishinposha, 2006, in Japanese), *The History of Korean Online Game Industry Development* (Seoul National University Press, 2006, in Korean), and *Economic Analysis on Game Industry — Structure and Business Strategy on Game Software Industry* (Toyokeizaishinposha, 2003, in Japanese).

He was also written many academic papers such as *The influence of social self-efficacy on online game satisfaction* (2007), *Organizational behavior of established firms to a disruptive innovation* (2006), and *The comparative study of online game user's attribute through path dependency* (2004). His paper entitled *Organizational design for new product architecture development* (2006) earned the best paper award from The Korea Society for Innovation Management & Economics, while another paper of his titled *The analysis of learning effects using online game community* (2005) won the best paper award from Korea Academic Society of Gaming.

Chapter 1

Innovations in the Game Industry: Online Games Versus Offline Games

1. The impact of online games: Turnover of Korean game industry against American and Japanese competition

Online gaming has been taking the world by storm since the 1990's, and Korean developers have been among the leaders. While Korean games are successfully exported to China, Japan, U.S. and other Asian countries, highly-developed manufacturing nations such as the U.S. Japan and several countries in Europe struggle to develop an identity in the online gaming space. The spectacular growth of online gaming marks a new chapter in the history of game industry.

Since the so-called "national awakening" in the 1960's, Korea strives to industrialize, modernize its economy, and compete on even footing with other advanced manufacturing nations. In fields such as electronics and automobiles, Korean companies imported technology and production techniques and worked to improve upon them, competing primarily on price with manufacturers from other countries. Now, Korean companies create some of the most competitive and advanced products in their industries, such as semiconductors, LCDs, and mobile phones, all based on the "import-and-improve" technique developed in the 1960's.

The online gaming industry stands apart in Korea. In gaming, Korea is now the market leader rather than the follower, both in terms of production and distribution. The first major online game, Ultima

Online, was American in origin but was commercialized and marketed by Koreans. Online gaming is a field where Korea no longer is forced to compete on price, but on technology and service. Currently, the Korean game industry maintains about a two year lead in technological innovation over its Chinese, Japanese, and American competitors (Wi, 2006a; 2006b).

Over half of the Chinese, Japanese, and East Asian gaming markets are made up of Korean games. Their spectacular success has attracted major game developers such as Sony Computer Entertainment and Microsoft, which are trying to build affiliations with Korean producers. This is significant in the development of Korean industry, as this is the first time Korean companies represent the pioneer in a major market.

Korean games usually operate on a different business model compared to other online games. Games created by American companies (such as Everquest and Ultima Online) use what is called a "selective monthly payment" model, in which users pay a monthly subscription fee for access to the game. Korean games generally operate via a system of micro transactions, where players have free access to the game but can purchase power-ups and aesthetic items for their characters.

One difficulty online gaming faces is that the advancement of the industry has outpaced the development of high-speed Internet services in many areas. Most new games require broadband Internet service to play; however, in America the ADSL network does not provide the capacity needed to play the newest games, regulating online gaming to a niche market. The American online gaming industry of roughly $700 million is overwhelmed with web games and the like, as most Americans do not have the internet services required to play cutting-edge games online. Japan faces similar issues, as the lack of available content and difficulties handling payment processing have stalled the development of its online gaming community.

The above explanations are reflected in Figure 1.1. Japan's online gaming community makes up about 1% of the entire industry, whereas Korean online games represent a whopping 62% of the total market.

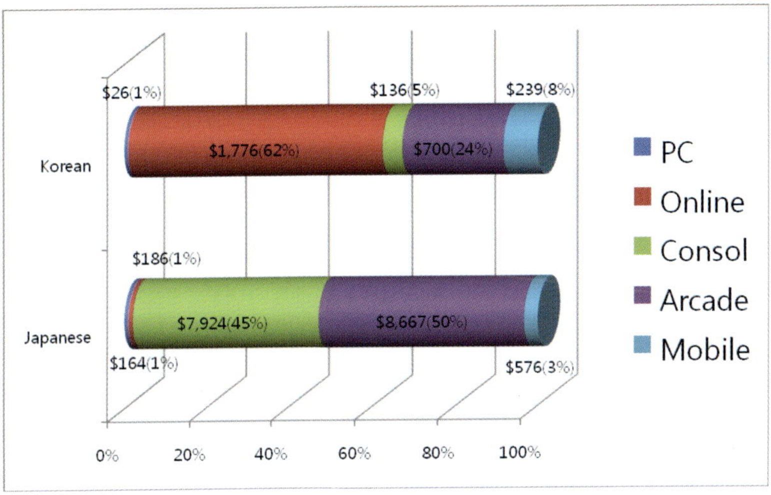

Figure 1.1 Size of online game markets in Korea and Japan
(Unit: 1 million dollar)
Source: 2007 The Rise of Korean Games

The Korean online gaming market has been experiencing explosive growth. As Figure 1.2 shows, the game industry began tracking market performance in 1998, with revenue of 20 million dollars in 1999. In 2000, the market had exploded to 1.9 billion dollars continuing to grow another 40% in 2001. Long-term projections predict an annual growth rate of 20% past 2007.

These numbers are even more impressive when considered in light of their exclusion of sales from third-party facilities such as internet cafes. In 2008, sales from Korea's roughly 20,000 internet cafes are predicted to reach 2.5 billion dollars.

The Japanese shrinking gaming market is a stark contrast, compared to the amazing growth rate of the Korean online gaming market.

2. The potential of online games

Online gaming has also seen explosive growth in China, Taiwan and many other Asian nations. The Korean market skyrocketed to $20 billion in just 10 years, driven by the launch of the immensely popular game

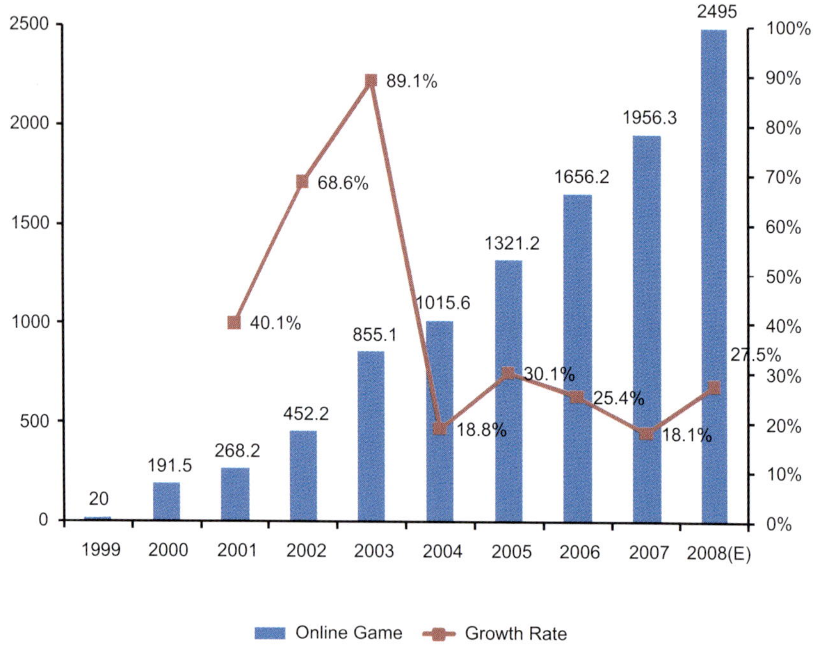

Figure 1.2 Development of the Korean online game industry (Unit: 1 million dollar)

Source: 2007 The Rise of Korean Games

Lineage in 1998. The Chinese market is currently experiencing similar growth, driven by Korean-developed games such as Legend of Mir II and B&B II, which have recently recorded over 700,000 players simultaneously logged into their games. The success of online gaming in Asia is far beyond anything seen thus far in America and Europe.

America and Japan are far behind Korea in the efficiency of their development process. Japan, long known for its gaming industry, is where many of the top console gaming systems were developed, such as the Playstation 3 and the Nintendo Wii. Japan is also home to many eminent console game developers such as Square Enix, Konami and Namco Bandai.

Japanese and American game companies face huge difficulties competing against Korean companies in developing games due to their massive development processes. They often spend many years and massive amounts of money on a title, only to see it reach a comparatively small market. A prime example is Square Enix's Final Fantasy 11, which cost around $100 million to develop and host. Koei's The Ambition of Nobunaga Online has spent over $10 million in development alone. Comparing those costs to the $3 to $5 million spent on an average Korean 3D online game, the inadequacy in Japanese game development and server hosting techniques are clear. American game developers face the same difficulties in their development processes, with costs similar to those of Japanese companies.

This disparity in development costs has led American and Japanese companies to be very interested in the Korean online game model. These companies hope to learn from and emulate the success of Korean developers. Mark Berner, the senior technical director of America Online, says the following:

> "I'm profoundly interested to see how casual online games have so spectacularly succeeded in Korea. And I believe that there's a good chance of importing those games to America, too. Of course, the cultural differences between the two worlds would have to be balanced first. But come to think of it, I see some positive prospects of such favorable Korean games being equally prosperous in American market as well."

As previously mentioned, Korean online games usually operate with a different business model from American and Japanese games. They feature cash shops, where players can essentially invest real money in their characters, making them more powerful. This creates a "pseudo-fund" of cyber money, online currency interchangeable amongst online gamers. This is an interesting type of capitalization that serves no real function as soon as the computer is turned off.

Yet, the value of these pseudo-funds is very real to online gamers. This cyber money identifies an individual and upgrades their status in their online game's society. In such a community, one is commonly judged by the strength of one's character; thus players are eager to invest in improving their characters. There is even a Japanese company whose sole product is online avatars for gamers. American game companies are learning about this business model and have begun to express interest in the Korean cash shop system. Jason Bell, vice president of American game developer Turbine, said:

> "I'm very intrigued by the profitable online game items in the American gamers' market .... users are willing to spend money on things like dressing or changing a hair-do of a cyber avatar character. I look forward to learning more about the Korean game corporate strategies as our broadband expands, soon to enlarge the online game community accordingly."

The actual online gaming industry is far larger than just the games themselves. Businesses supporting mobile accounts for payment processing are completely independent from game developers, and such businesses are successful primarily due to online gaming. The vast majority of their transactions come from teenagers who are unable to obtain credit cards, and use these payment processing services to pay for their gaming.

Internet cafes also benefit greatly from the growth of online gaming. Internet cafes offer high-speed broadband access for optimal gaming, and Internet cafes are springing up in other Asian countries to serve gamers who are unable to get high-speed access at home.

Online gaming is even spreading to sports. The concept of "E-health" is a new attempt to combine gaming and sports. This is being tested by a Korean company that has developed a system in which the game character's earned experience points are increased based on the distance the player runs on a treadmill which is hooked up to the computer. This experiment may be an attempt to resolve the "gaming addiction" problem that some players experience.

Online gaming is here to stay, and it is spreading like wildfire. All console games will soon feature online components. In time, every video game (whether played in an arcade, on a console at home, on a personal computer, or on a cellular phone) will be played online to add more content and connect gamers who share similar interests.

The present online gaming market is just an initial stage, a test of how the Internet will eventually influence the gaming industry as a whole.

Many people regard online gaming as a hobby. Some even call it a dangerous addiction. But clearly the impact of online gaming is being felt in today's societies. How did it come to hold such influence, and how will the differing cultures from which players come affect how they interact online? In this book, we shall explore the growth and the impact of the online gaming industry in detail.

3. Innovations of online games

While the world continues to witness an explosive expansion of online games with millions of players logging in from almost every country on the planet, the entertainment industry and other businesses position themselves to take advantage of this exciting growth. Relatively few people understand that the rate of expansion could be even greater were it not for an inherent tension between offline games and online games. To understand how this is the case, we should look at the fundamental differences between offline and online games.

In the past, all video games were offline. Users played games on ROM cartridges or CD-ROMs via game consoles, PCs and TVs. For players, this meant either playing alone or with a few other people in the same real space using the same, single piece of equipment. In the offline game business, software makers developed the game and sold it, usually through traditional retail channels. This business model has been dominant in the offline game industry for over 20 years.

The mid-1990's, saw the emergence of online gaming. These online games involved users connecting to servers via game consoles, PC's and communication networks. Players no longer needed to be in the same physical space or use the same machine or console in order to play a game, and over time they were able to play with more and more people. Using their own machines and connecting from remote locations around the world, they were able to log in to the same game simultaneously. Today, players interact with hundreds or even thousands of other players within online game worlds (Wi, 2006a).

It is crucial to understand that the creation of online games was induced by disruptive technology and business model. Technology developed to make offline games more enjoyable for players by allowing them to play with others has led to the tension in gaming we see today. When a disruptive innovation so different from previous technology and existing business models is generated, even leading companies often fail to adapt to the newly created environment. When technology or the very structure of the product undergoes enormous changes in a short period of time, such as the changes involved in going from offline to online gaming, management resources accumulated by the corporation can turn into obstacles and hinder the organization's ability to adapt. Christensen (1997) has described this phenomenon as "the innovator's dilemma."

To best appreciate the dilemma offline game makers faced in the past and continue to face in the present vis-à-vis online games; as well as to understand the challenges online games have to contend with and continue to have to confront because of offline games, we can compare and contrast the two forms with respect to several key differences. The differences on which I will focus are as follows: 1) revenue model; 2) product attributes; 3) development process; 4) distribution channel; 5) communication; 6) game style; and 7) immersion.

1) *Revenue model*

The first difference between online and offline games involves revenue and how it is generated. With offline games, the developers' source of revenue is via the retail sales of their game. However, for online games the main source of revenue is the recurring fee users pay for connecting to the game server. Let us look at how these sources compare.

Offline game developers acquire revenue through software sales. Prospective players must purchase the console and the software in order to play the game. As shown in Figure 1.3, a piece of game software that retails for $58 returns a margin of $18 to the retailer and $10 to the wholesaler, with $10 going into paying for production and any royalties. The developer is ultimately left with a profit of $20.

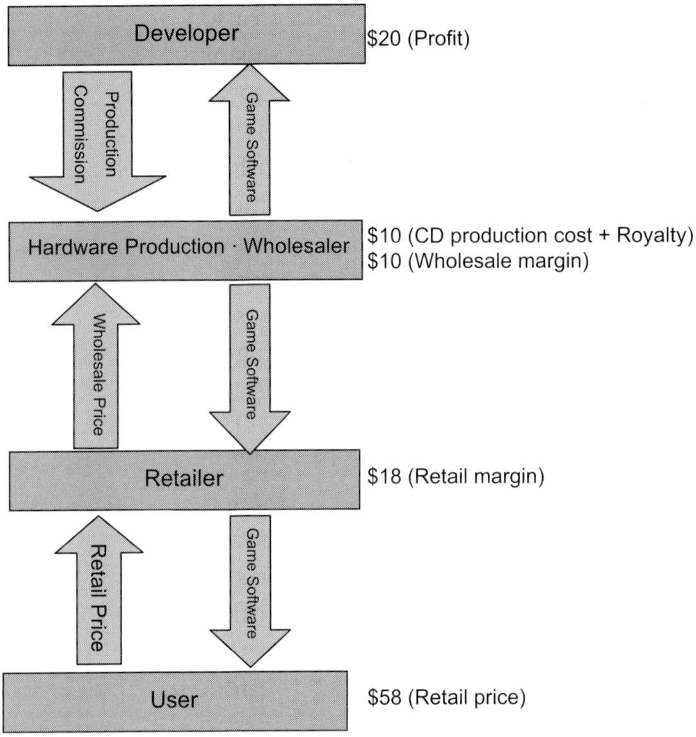

Figure 1.3 Revenue distribution structure of offline games

If development, advertisement and sales cost a total of $2 billion for a title, 100,000 units of the software must to be sold to retrieve that cost. Offline games require only a one-time payment: when the consumer purchases the software. Thus, with offline games, additional costs cannot be retrieved aside from through the profit gained from software sales.

The profitability of offline games rests solely on how many game packages can be sold. For the majority of offline game software packages, sales peak with release and decline with time, adhering to a right-descending pattern. Seen as a whole, most of the total sales volume is concentrated in the extremely short period immediately following release, in what is known as a distinctly characteristic "early concentration" sales pattern (Shintaku *et al.*, 2003). This makes the

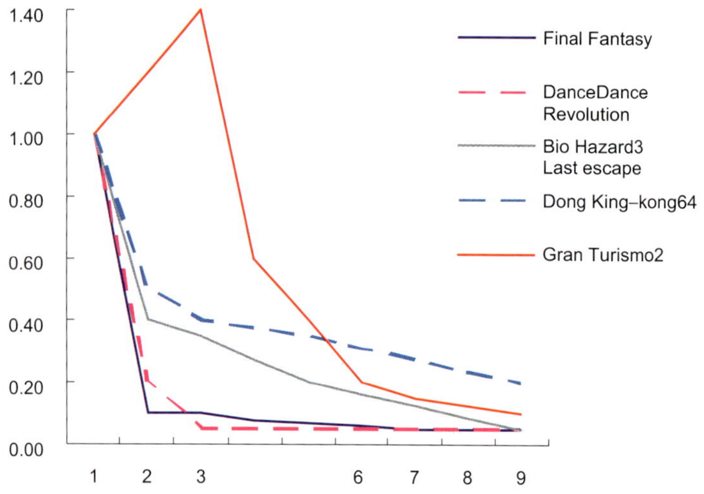

Title	Maker	Platform	Releasing date	Total number of sold units	Number of sold units in releasing week	Ratio of selling in relasing week (B/A)
Final Fantasy	Square	PS	1999.2.11	3,470,590	2,502,859	72.10%
DanceDance Revolution	Konami	PS	1999.4.10	1,007,811	261,616	26.00%
Bio Hazard3 Last escape	Capcom	PS	1999.9.22	1,379,329	1,002,614	72.70%
Dongki kong64	Nintendo	N64	1999.12.10	1,089,825	189,669	17.40%
Gran Turismo2	SCE	PS	1999.12.11	1,764,922	815,430	46.20%

Figure 1.4 Sales pattern with first-week sales set as 1

1) Abbreviations of platform are as follows. PS = PlayStation, SS = SegaSaturn, N64 = Nintendo64, DC = DreamCast, PS2 = PlayStation2. These are used in the following figures.

Source: Shintaku *et al.* (2003), p. 151

decision of how many game units to produce and release in the first week of sales critical (see Figure 1.4).

Analysis of the game software sales data between 1997 and 2000 indicates that most products adhere to the early concentration sales pattern and the degree of early concentration increases annually. Simply put, practically none of the games that performed poorly in the first week ever recovered. Also, given the data, we can understand why a product is marketed intensely immediately before and after its release.

Consider the case of Final Fantasy 8, developed by Square Enix. While total sales reached 3.47 million units, 2.5 million of those were in the first week after release, producing a first-week sales concentration of 72%. It is this sort of performance that drives offline game developers to concentrate their marketing efforts immediately before release in order to maximize sales.

But with online games, payment for rendering services instead of payment for software is the main source of revenue. Users of online games are required to pay a connection fee to access and play the game. These fees range between $20 and $30 (in the case of Lineage, $28.5). With the partial pay model of item sales, users pay for their server connection through making in-game item purchases. If a server connection fee is not paid consistently on a monthly basis or in the form of item purchases, users of online games are typically unable to connect to the server or cannot play the game smoothly or to their full satisfaction. Of course, online games require the game to be installed, usually by downloading the client from the Internet. But this does not necessarily constitute a sale; it is merely a process of exchanging information.[1] With online games, once users are secured, a continuous stream of revenue can be generated. The strength of communities formed within the game is one of the determining factors of success for online games, and this leads us to the next key difference.

---

[1] 'Ever Quest 2,' a recently released 3D graphic online game, requires 4.3 gigabytes of software to be installed in the PC. This caused daylong downloads during the beta phase when prospective users inundated the game.

2) *Product attributes*

Offline and online games differ with respect to product attributes. Product attributes refer to endemic attributes that can be identified and evaluated by users (Lancaster, 1991). Before proceeding to discuss the product attributes of offline and online games and the major differences between them, let me elaborate on what product attributes are and explain how they are important.

A notebook PC, has two critical product attributes. A notebook PC is typically evaluated according to 1) how little it weighs and 2) how efficiently it processes information. These two product attributes are important in notebook PCs, but their relationship is inversely proportional. The lighter the laptop is, the less efficient it will be. To shed weight, CPU efficiency has to be scaled down and generated heat has to be reduced, as does the hard disk capacity. Accordingly, laptop developers must choose between championing lightweight machines or information processing ability, or attempt to balance the two. Consumers manage their priorities with respect to machine weight and information processing ability, and they make their purchases accordingly (Wi, 2004, 2006b).

Differences in product attributes exist between offline and online games and the central tension arises from the fact that an online game is a type of modified offline game architecture with community identity added to the gaming experience (Wi and Nojima, 2003). The gaming experience is a general term referring to what users experience and evaluate with respect to graphics, sound, story structure, character design and other categories that fall under what it is like to play the game. In other words, the aesthetic and functional satisfaction, or lack thereof, users obtain from the game is referred to as gameplay. On the flip side, community is a product attribute particular to online games that creates a sense of solidarity, camaraderie and satisfaction amongst users via communication, competition and other forms of interaction. Many user-communities such as guilds, clans and alliances have been formed in online games. These communities facilitate the gameplay of new members, and strengthening a sense of unity for community members is critical to generating quality extended gameplay and thus retaining players over the long term.

Offline games usually involve one player vying against a preset program while online games feature a massive number of users engaged in mutual warfare or cooperation. Accordingly, offline and online game preferences vary. Kenji Matsubara, CEO of Koei puts it as follows.

> A majority of offline gamers play alone, slowly and deliberately. In our "Romance of Three Kingdoms," individual players ponder the game at work or at home, devising strategies alone. But it is different with online gamers. They engage in speedy gameplay via communicating (chatting) with other users, rapidly increasing their experience rate while hunting monsters.

Hence, offline gamers usually engage in individual play while online gamers cooperate and compete with other users via communities.

Offline games strive to offer high-quality sound, graphics and plot, and those gameplay elements are often their main focus. One of the goals of the Japanese developer Square-Enix is to create outstanding graphics no matter how much development cost it may entail.

However, that is not the case with online games. Gameplay elements such as graphics and sound must be balanced with community. Unlike offline games, community is critical in online games and so chat and other enabling functions are offered. The developer of an online game must therefore strike a balance between pursuing excellent graphics, sound, and plot with other needs such as supporting user communities. Online game developers support user communities in various ways by creating communication tools such as chat windows, building user events and Web sites, and holding offline meetings.

Originally, guilds were primarily formed in order to collectively participate in sieges and hunts; however, guilds with communication objectives are on the rise. For example, in Lineage 2, one of the largest MMORPGs (Massively Multi-player Online Role Playing Game) in terms of active subscribers, diverse communities whose objectives are other than sieges have emerged. Within those communities, chats and debates on common hobbies and interests take place, and specialized groups such as fan clubs, married player groups, same-age player groups and groups of relatives have formed.

The sense of community is usually extremely weak in offline games, if not non-existent. Because of that, user management or service to players following release is of little, if any, importance. But with online games, continuous user management, especially with respect to community, is required after the service is launched. As the number of user communities increases, the value of the game increases. This leads to more players, and again greater value. A set of more players can result in a larger monthly user fee, which in turn translates to greater corporate profits. This increased revenue also enables the developer to invest in system upgrades, which increases the appeal of the game, and in turn induces a further surge in the number of users. A firm that has secured such a propitious cycle will be able to obtain millions of users.

3) *Development process*

The third difference between online games and offline games involves development as it relates to technology and process. Technology needs differ between offline and online games. Unlike with offline games, servers and network technology are essential for online games. Offline games do not require network technology as they involve an individual player vying against the game or multiple players waging battle against one another but playing on the same machine. Online games are composed of client programs which install on the user's PC and on the server (see Figure 1.5), while offline games are composed of only the client program on the user's machine or on a disc.

With online games, many players engage in gameplay via the server. Hence, technology addressing how to accommodate players on a restricted server and how to deal with an inundation of players logging onto the server at once are essential.

Consequently, server management, maintenance and repair costs are a formidable expense that only online game developers face.

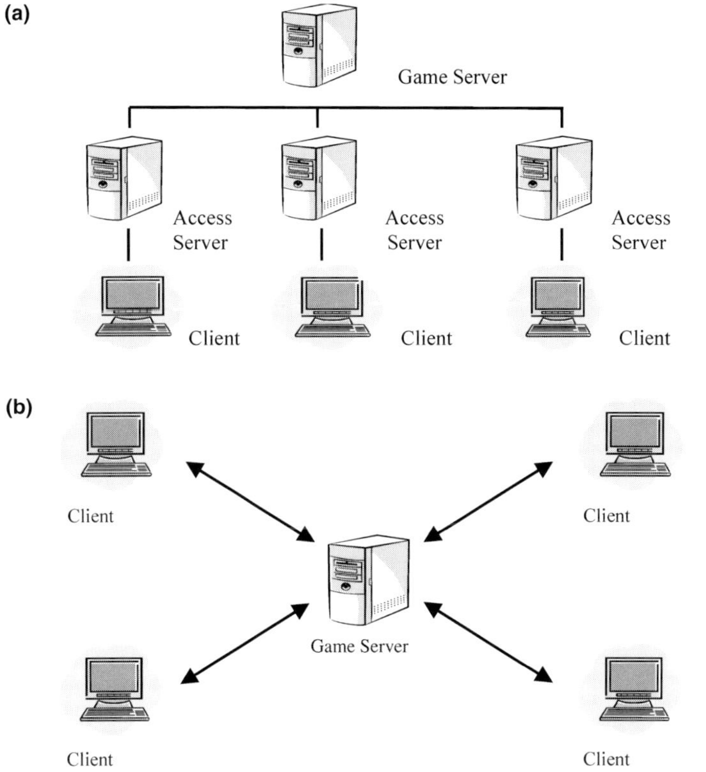

Figure 1.5 (a) Distributed server model (b) Client-server model

Jake Song, a former Executive Vice President of NCsoft, recounts an early server programming learning process.

> "I majored in computer science in college and picked up server technology by collecting text MUD open sources from the Internet and elsewhere. I obtained MUD sources and studied and improved them. I deemed myself an able programmer but it was my first time dabbling in network programming. By providing game services, I discovered many ways to increase server efficiency with repeated improvements. I took a network-programming course in college and it was a big help. Those who took graduate courses in server or network technology have an advantage."

This combination of server and network technologies required for online game development makes a smooth transition from creating offline to online games difficult for dominant offline game companies. For Japanese companies looking to expand beyond offline games, there are few engineers available who have developed games based on server and network technology. The lack of experienced developers to build server programs requires extensive trial-and-error, and consequently an enormous cost in online game development.

Another way we can appreciate the difference between offline and online games in terms of development is to consider the development of offline games as analogous to film production. Once a film reel leaves the editing room and is transported to the theater, no more revisions are possible. With offline games as well, once bugs are removed in the final stage and CD-ROM production is under way, no more revisions can be made. If a severely detrimental bug is detected after the offline game's release, the only remedy is to recall it. But it is a different story for online games. Even after an online game is released to the public, continuous content revision is possible.

Online games can be revised in various ways with respect to a number of post-release needs. For example, character ability balancing is critical in online games. If one type of character is especially strong, word spreads among users that this type of character is easy to do well with, that it has an advantage over other characters. Soon, players swarm to play that type of character. When something like this overpowered character class occurs, the GMs (Game Masters) who monitor the game report it to the developer, who can implement a patch to address character balance. The same applies to controlling the volume of virtual items and the game's economy.

Players tend to concentrate at sites where it is easy to acquire specific items or in-game money. If it is determined that a game item being traded — a powerful sword, for instance — is too easily found or purchased, the developer lowers the acquisition rate. That way, the supply decreases and prices increase, and not as many players are able to wield such a powerful sword. In online games, users are constantly observed, enabling developers to address their needs via system modifications in real time. Returning to the film analogy, it is almost as if

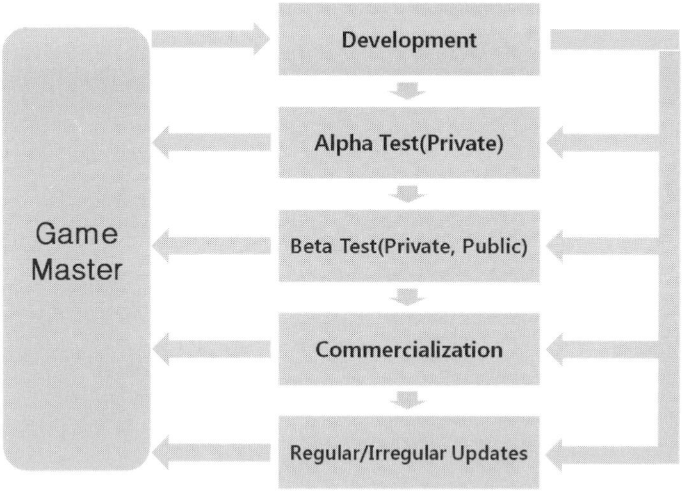

Figure 1.6 Online game development and update process

a film could be edited, enhanced or expanded while being shown in a movie theater.

Online games evolve through mutual interaction with users, and this interaction typically proceeds through stages. Figures 1.6 and 1.7 refer to online game development and the update process. After an offline game is developed and released, the development team is disbanded and development-related tasks are terminated. Online games, however, go much further.

After the basic game has been built, a point at which an offline game would exit development, the online game enters a closed (private) alpha test phase. At this stage, the development team is not disbanded but supplemented with additional developers to prepare for revisions. The closed alpha phase entails testing the game and detecting bugs with a limited user pool, the number of testers ranging from the dozens to the hundreds. The game is not available to the public at this stage.

The second stage is the closed beta test. The beta is different from the alpha test phase in that the game has been further developed and more users are added to the test pool. For high-profile games that enter the beta test phase, participants are often selected from drawings.

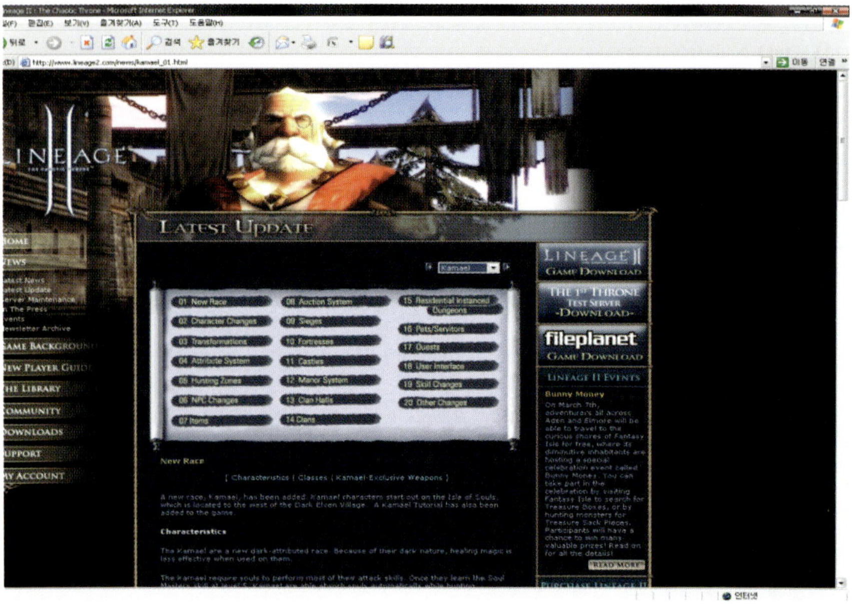

Figure 1.7 Lineage 2 update instructions

Source: <http://www.lineage2.com/news/kamael_01.html>

Such selection processes are extremely critical marketing decisions, as many of the gamers participating in these beta test phases have the power to sway public opinion. As such, it is important to take the judgments of private test participants very seriously and possibly make revisions reflecting their preferences.

At the third stage, the online game advances to the open beta test phase. Many more users are invited. This large pool of users actually tests the game under the same conditions as when it will be offered commercially. The open beta phase is critical because some of those who participate in the open beta phase will remain when the game goes commercial.

It is standard for about 10% of the users from an open beta test to remain when an online game goes commercial. The transition from the beta phase to the commercial phase is determined to be a success if more than 10% of the users remain. In a case where 1 million users access an online game in the open beta phase, the goal would be to have at least 100,000 players remain. If 100,000 users participate, 10,000 are likely to remain.

When an access fee is charged, there is a substantial difference between 10,000 and 100,000 users remaining. In order to make 10,000 swell to 100,000 after the game's launch, enormous amounts of money must be spent on marketing. But if 100,000 remain and the game's bottom line is immediately in the black, then the developer has achieved an excellent head start. This is why a developer will often try to have as many players as possible participate in an open beta test.

The final stage is commercialization. In the commercialization stage users are charged a fee to play the game. Set monthly fees range from $20 to $30. At this stage, users who accessed the game in the open beta stage must decide whether to remain or to leave. For the game developer, deciding on the duration of the open beta phase is critical. A lengthy open beta raises the probability of communities forming and the members of those communities wishing to stay connected with one another, but the cost to the developer increases. In contrast, a short beta period cuts costs but the probability of users remaining drops. Consequently, aptly adjusting these two opposing factors becomes a key component to the game maker's strategy for when to commercialize.

Throughout all phases, GMs (Game Masters) perform a vital function. GMs absorb user comments and requests and relay them to the development team in the form of feedback during the alpha, both beta, and commercial phases. Developers repair bugs and execute updates based on that feedback. Of course, the function of the Game Master is non-existent in offline games.

With online games, multiple complex phases exist across development wherein constant interaction with users is possible, thus enabling perpetual content revision. If the reception is tepid in the alpha test, a major overhaul can be performed in the beta phase so as to try and produce an improved reaction that will hopefully lead to success in the commercialization phase. This ability to explore user reactions and apply changes based on those reactions is a critical difference between online and offline games.

4) *Distribution channel*

The fourth difference between online and offline games is the manner of distribution.

With offline games, the game console manufacturer is routinely commissioned to manufacture CD-ROMs containing the game software, which are sold via retail and wholesale vendors to the consumer (see Figure 1.8). CD-ROM packages containing the software for game consoles are sold via two distribution trajectories. For one, the package is manufactured by the software producer and sold via retail and wholesale outlets to the consumer. Moreover, the package is manufactured according to the above method and reaches the user by way of the software developer directly, such as via website or catalog sales.

In cases like those discussed above, the software company typically develops a game autonomously with the game console selected as the foundation for development. As game consoles are stable for a set period without much changes to their features, the developer can concentrate exclusively on software development. For offline games, software developers and manufacturers are disparate with respect to development and distribution. With such a clear division of labor and vested interests, game software developers need only to focus on developing the software.

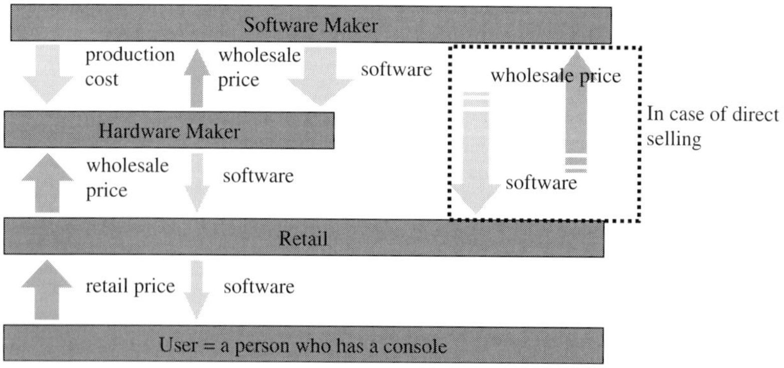

Figure 1.8 Distribution structure of offline games

Source: Wi (2003), p. 317

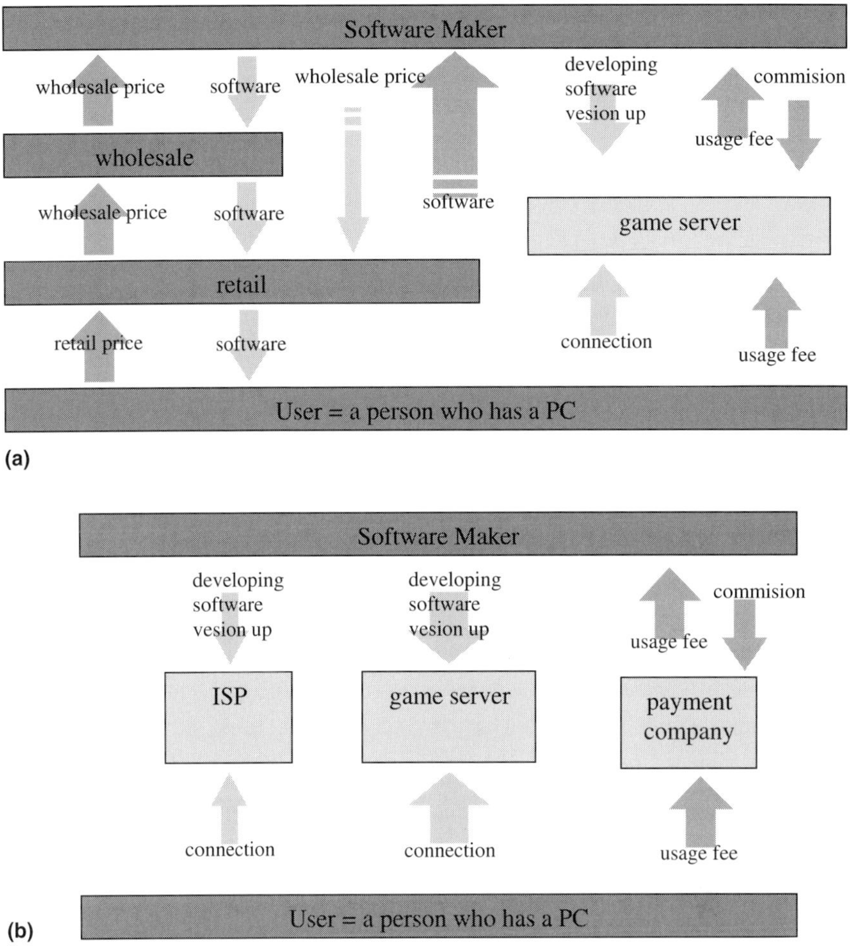

Figure 1.9 (a) Online game: Type 1 (b) Online game: Type 2

Source: Wi (2003), p. 317

For the sake of this discussion, present online games can generally be classified into Type 1 and Type 2 of Figure 1.9. Two business models exist in the online game industry, one for each game category. Type 1 is popular in the U.S. market. MMORPGs (Massively Multi-player Online Role Playing Games) such as Everquest and WoW (World of Warcraft) are typical online games using this type. Game developers sell CD-ROM type online game through an offline distribution channel.

As users buy and install the game, they have to access a game server and start playing a game. In this case, users have to pay twice for playing online game, CD-ROM package and monthly fee.

Web-based board games such as chess, mahjong and poker also belong to this category. With online simulation games the server's role does not extend beyond that of the middleman linking users in one-on-one or larger-scale gaming.

Web-based games are relatively simple games requiring little disk space and can be produced with little difficulty. Simple game development can require as little as one development designer and two programmers, and one designer may design multiple games. Users of Web games go online, access game Web sites and download clients that are generally less than 5 MB in size. These games are played when users enter one of multiple game rooms built on the hosting website.

In the distribution model of online simulation games reflected in Figure 1.9(a), the package sales method by way of wholesalers and retailers and the direct sales method of the software developer installing the server and directly offering the game to users co-exist. Online simulation games create revenue through package sales or game item sales, and software developers for these kinds of games rarely charge monthly fees. In fact, most Web-based board game revenue is created by additional paid services and avatar sales.

Figure 1.9(b) is a distribution model of Type 2. Type 2 is a game wherein warfare is waged by a massive number of players who connect to the server simultaneously and engage in competition and cooperation. Some leading MMORPGs reviewed for this study are the following: NCsoft's Lineage (Korea); and Square Enix's Final Fantasy XI (Japan).

With MMORPGs, frequent content upgrades are essential to ward off user boredom, which could lead to subscription cancellation. In other words, to keep current users from growing bored as well as to draw new users, performing incessant revisions that inject new, appealing content is imperative. NCsoft performs minor daily revisions and major updates every six months. Lineage first began with Episode 1 of the Small Island; the game currently boasts more than 20 episodes.

To develop an MMORPG, anywhere from hundreds of thousands of dollars to millions of dollars is required, and the development period typically lasts at least two years. A team of 70–80 individuals developed Lineage 2, with 3 content producers, about fifteen programmers, and well over 50 graphic and sound designers.

The defining attribute of the Type 2 model is the disappearance of the middleman. Either the developer or the publisher possesses and operates game servers in-house. The servers are directly owned and operated by the developer, or the servers are placed on IDC and the management of the servers is contracted out. In offline games the distributor was a separate entity, but in online games the developers have personally taken on distribution tasks. Because of this, in countries like the U.S. where offline games are dominant, a resistance of retailers to online games exists. These middlemen are aware that the distribution model of online games has excluded them. To overcome this resistance, when NCsoft was distributing Lineage 2 in the U.S., they signed a contract precluding online downloads and sold the game via offline distribution channels only.

While the revenue for offline games is dispersed as it passes through retailers and wholesalers, with online games the developer receives a direct payment by taking advantage of such methods as Payment Gateway (PG) when the user has chosen to pay by cellphone. In this process, the developer pays a fee to PG. In addition to the monthly fee the users pay, the developer directly manages distribution and payment. To collect server connection fees online, building a network of payment methods using landline and wireless phones, Web money and credit cards is essential, as is maintaining a high-speed infrastructure to enable the massive exchange of information between users and the company.

5) *Communication*

Community identity is the single most important product attribute of online games, and what most distinguishes online from offline games. Community identity is most profoundly enabled and reinforced through communication between users. User communication is rare if not non-existent in offline games, whereas in online games it plays a vital role.

When novice players first access an online game, advice and help from experienced players can be very valuable, not just to the new player but to the community and game as a whole. Such advice and support contributes to motivating the novice player to remain in the game. Thus it is in the developers' interest that experienced players want to assist new players. For this reason, developers often systematize fellow-player or master-student relationships in their games. In "Legend of Mir 3," if a master-student relationship is established in the game, a valuable ring is awarded to the two players. Accordingly, among the reasons for continued play as reflected in Figure 1.10, more online game users replied "because of in-game friends" than did those of offline games.

In raising the level of characters created within the game, party (group) play is more conducive than individual play and communication plays a key role. In MMORPGs, character classes like warriors, mages and archers are assigned discrete roles. When players group together in order to benefit from one another's strengths and abilities — for example, in a monster hunt the mage fortifies the warrior's strength with spells — they can more easily achieve their goals or even accomplish tasks impossible to them if they played alone. The chat window on the game screen is a tool that enables communication for team play so that players can coordinate their efforts.

On the contrary, in Dead or Alive Xtreme Beach Volleyball for the Xbox, two or four players can play using the same console. Most offline games involve players vying against only the game program in a single-player mode. But even in the case where players can share a console and play together, there is little to no need for offline gamers to use communication tools within the game's software (they are usually in the same physical space and can communicate to one another directly). As illustrated in the Figure 1.8, the "degree of engaging in gameplay in conjunction with fellow players" tends to be higher for offline games. The degree of cooperation of identical online gameplay is higher for those who primarily play online games. Also, in

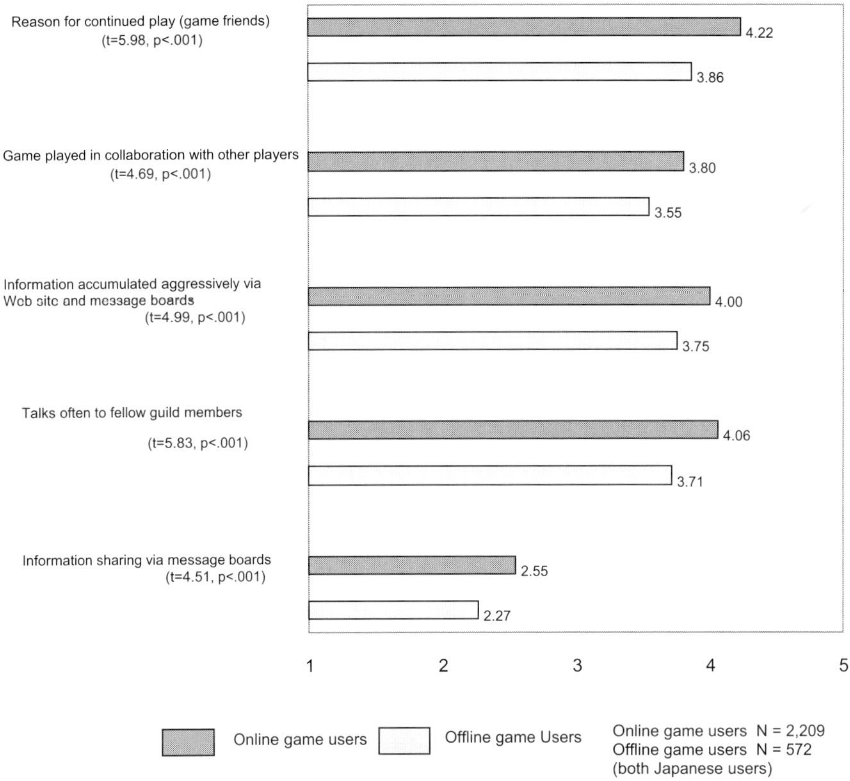

Figure 1.10 Inter-user communication[2]

"communication frequency," "aggressiveness of acquiring game information" and "sharing game information," online gamers scored higher than did offline gamers. As reflected in Figure 1.10, the "frequency of players talking to fellow players" is higher for online gamers than it is for offline gamers.

For players of online games, collecting and distributing game information is critical. Possessing information on where valuable resources can be discovered on the game map, what valuable items

[2] 'Online game users' and 'offline game users' are divided by platform preference. Through both user group is playing PC online games, 'offline game users' are playing games more on the base of console platform.

can be obtained from which monsters inhabiting which locations, and which guilds are beneficial have an enormous influence on gameplay. Accordingly, online game users aggressively compile game information and share it with others. On occasion, when an online game user posts information on the game Web site, other players offer in-game help, their own information or items in return.

But it is different with offline games. Information needed to play the game is distributed among users, but it does not determine player ranking or influence in-game power relationships. Possessing game information may be a cause to boast to friends, but lack of it does not pose a detrimental problem in gameplay as it does in online games. Accordingly, as reflected in Figure 1.10, for the two indicators "information accumulated aggressively via the Web site or message boards" and "information sharing via message boards," online game players scored higher than offline game players.

6) *Game style*

Differences exist in play styles between online game users and offline game users:

First, the concentration of online game users fared higher than that of offline game users vis-à-vis whether they are playing only one or multiple games simultaneously. More offline game users are playing multiple games simultaneously than are online game users. The issue of game concentration is linked to fees and the rate of change within the game. Online games charge a fee, and the fee is charged even while one is not playing. Consequently, for online game users to play multiple games at once, the cost would multiply accordingly. As such, online game users routinely exhaust a game before moving on to a different one.

But with offline games, the consumer incurs no additional cost after purchasing the game's software. Hence, even if they play multiple games they have purchased simultaneously, no additional cost is generated. Consequently, the game concentration of offline game users is higher than that of online game users.

The rate of change within an online game is another determining factor for game concentration. Offline game players can pause and resume gameplay; however, online games cannot be paused for individual players. If a player leaves the gameworld, other players continue to play, accruing experience rate[3] and strengthening their personal relationships with fellow players. When the player logs into the game again, the game will have undergone substantial changes, especially if much time has passed. From this perspective, an online game resembles a persistent world of its own rather than anything like a simple offline game. Truly, such online games are virtual worlds undergoing constant change.

In most online games a strong correlation exists between the duration of gameplay and character level. The character level symbolizes the degree of power that a character possesses within the game and serves as an important motive for online game users to invest more time in the game. As reflected in Figure 1.11, whereas the average duration of gameplay for online game users is 3.8 hours/day on weekdays and 6.5 hours/day on weekends, offline game users play 1 hour/day less on weekends and 30 minutes/day less on weekdays.

Heavy users of online games log on to the game every day in order to keep tabs on the game's progress and communicate with in-game friends. The number of players connected to the game peaks at night for most MMORPGs, though there is a sharp surge in hits from 7 a.m. to 8 a.m. when players check in on the events that have unfolded overnight before they leave for school or work.

The community attribute of online games is also a factor that contributes to extending gameplay duration. Association with in-game friends extends over and above the core gameplay, to taking walks within the gameworld or having conversations on mutual real life interests. In Lineage 2, a guild for married players exists. The only requirement to joining this guild is that players be married, and

---

[3] Experience rate means character's level inside a game. Characters with higher level get a power than that of lower level.

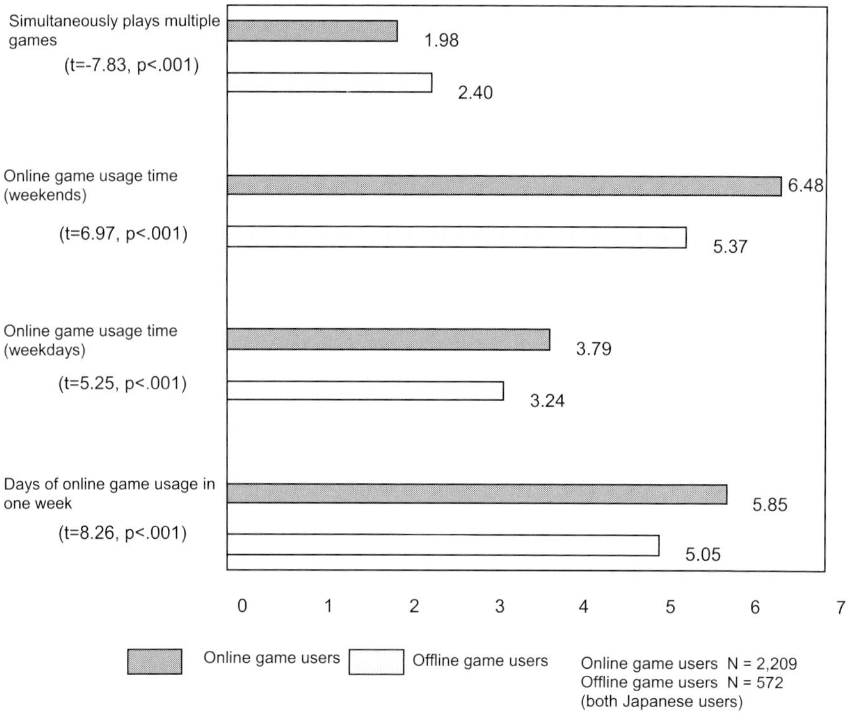

Figure 1.11 Online game styles

topics of conversation range from family issues to children's education. The motivation to level up one's character in this guild is relatively weak. In Lineage 2, members of this guild can be spotted taking walks and holding conversations sitting in groups. Such communication extends gameplay duration.

Online game makers must contend with users who are so fixated on character level that they quickly exhaust core gameplay. When players have completed all the quests or activities in the game or have reached the highest possible level, users are sometimes no longer motivated to play on. This is a critical issue for game companies who offer services in developing countries whose users are routinely extremely fixated on maximizing their character's level. The company must continue to perform updates and patches suitable for high-level users in order to keep them playing, and this typically results in paying for additional development personnel.

For offline game users, game duration is solely comprised of actual gameplay. It is rare for them to suspend playing and talk to other users because for most offline game users how fast they complete the game is often the primary, if not the only, motive for playing.

7) *Immersion*

The final key difference between online and offline games to be discussed in this chapter is immersion. The fact is already pointed out that the duration of gameplay is longer for online gamers than it is for offline gamers. Duration is usually proportional to the degree of immersion; hence, we can conclude that online game users are deeply immersed in the game.

The difference in the degree of immersion is palpable when compared to that of offline game users. More online game users answered in the affirmative to the whether or not the following applied to them: "I think of the game even when not playing", "I play the game in my free time," or "I feel depressed when I don't play and that depression dissolves if I play."

Online games comprise virtual societies which continue to change even after users stop playing. When the user returns to the real world, diverse incidents and events continue to unfold in the game world. For example, take the case of a siege undertaken to secure a castle. Seizing a castle is a very exciting event, and when a user is forced to withdraw from the game for personal reasons while engaging in battle with in-game friends, that user will inevitably be preoccupied with the game upon returning to the real world.[4]

Such powerful absorption of online games produces game addicts who are unable to restrain themselves from playing compulsively.

---

[4] Some researchers are thinking that online game's immersion can be used in the educational field. They already tested online game whether it is useful as an educational tool (Wi and Oh, 2007; 2006a; 2006b; Wi and Won, 2006).

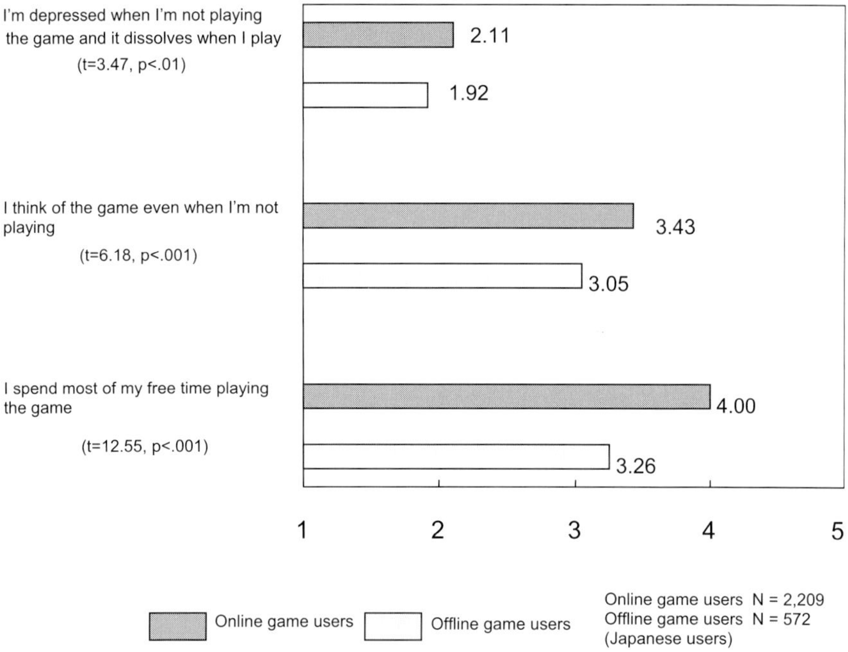

Figure 1.12 Immersion in the game

Below is a testimonial of a user whose absorption with an online game has led to neglecting offline human relations.

> I quit school one year ago. For one year I didn't venture out of the house even once a month. I only played my game at home. I've increasingly grown indifferent to world affairs and going out doesn't interest me anymore. I've grown so obese now that I don't even like to walk. If I walk for too long my back aches and I grow testy. I know this is no way to live but outside of the game I am apprehensive, nervous and fickle. I've even grown distant to my friends. I haven't made any new friends since I graduated from junior high.[5]

[5] Electronic Times, May 10, 2004.

The above individual assigns deeper meaning to online relations over those offline. Players sometimes derive a sense of achievement from accomplishing goals in online games, a feeling which eludes them in the real world.

For instance, teenagers who are hounded by feelings of inferiority induced by poor performance at school can achieve tremendous status in an online game. A player might become a guild leader or the lord of a castle with tens of thousands of subjects. When players become castle lords, they are able to wield enormous power in issues ranging from taxation to conducting battles. The game world provides an alternative environment wherein teenagers are able to achieve what they have been unable to in the real world. In some occasions, it goes so far as to replace the real world. These unique characteristics of online games contribute to boosting the degree of immersion of online gamers.

Understanding how online games developed from offline games, and how there is a tension between these two forms due to their differences, we are now ready to look more closely in the next chapter, at the arrival of online games in the world to get a better sense of what they are and how they operate with respect to business.

Chapter 2

Business Models and Corporate Strategy

There are several different proven methods to draw revenue from a successful online game. The right method must be chosen for each game, which is no easy task. Remember that, as previously discussed, online games differ from offline games in that offline games generate their revenue via one-time sales of the game software, whereas online games generate revenue via persistent payments made by users in order to play the game. Thus for an online game to be successful, it needs not only attract users initially, but it needs to keep them there long-term. The revenue models are very different. Offline games are essentially a retail industry, whereas online games are a service industry.

The two primary revenue models of online games are charging fees directly to individual players and charging fees to Internet cafés to allow them to provide the game. There are two types of pricing plans available: set-amount, where the user pays a monthly fee for unlimited access to the game; and set-volume, where the user prepays for a certain amount of time to play the game.

The set amount pricing plan is generally favored by Asian consumers, as they prefer a prepaid "pay as you go" plan, or a game that is free to play and sells items to generate revenue (known as the "partial pay" method).

1. Revenue models of online games

Online game fees are distinguished between two types: fees charged to individuals, and fees charged to Internet cafés. There are 5 main

Table 2.1 Online game price plans

Model		Defining attribute	Notes
Individual	Set-Amount–①	Fee paid for a set period	Prepaid
Individual	Set-Volume–②	Fee paid for a set usage duration	Prepaid
Individual	Partial Pay–③	Fee paid in the form of game item sales	Cash charged
Internet Café	Set-Amount–④	Fee paid for a set period	Prepaid
Internet Café	Set-Volume–⑤	Fee paid for a set usage duration	Prepaid

Table 2.2 Set-Amount price plan for individuals for World of Warcraft

Product	Price
7 days (1 week pass)	$8.80
30 days (1 month pass)	$25
90 days (3 months pass)	$60

Source: <http://www.worldofwarcraft.co.kr/billing/index.html>

revenue models within these two classifications, as denoted in Table 2.1.

1) *Individual Set-Amount plan* (①)

The set-amount plan allows a user to pay a monthly fee for unlimited access to the game. This plan benefits gamers who play one game regularly, but quickly becomes expensive if a player wishes to split their time between multiple games. The monthly fee is paid regardless of whether or not the account holder logs in to the game during the prepaid period. Set-amount plans generally offer discounts for purchasing longer subscriptions up front. World of Warcraft, a leading MMORPG offers three different set-amount plans for users to choose from.

Table 2.3 Set-Volume price plan for individuals for World of Warcraft

Product	Price
5 hours (300 minutes)	$4
30 hours (1,800 minutes)	$15

Source: <http://www.worldofwarcraft.co.kr/billing/price_fixed.html>

2) *Individual Set-Volume plan* (②)

The individual set-volume plan allows users to prepay for a specified amount of gameplay time. This plan is beneficial to gamers who wish to play many different games, not focusing their time on one game entirely. A player's prepaid time lasts until it is used, but if a player wishes to play one game heavily, this method generally ends up being more expensive than a set-amount plan. For a direct comparison of the two pricing plans in the same game, World of Warcraft's set-volume pricing options are listed above.

3) *Partial pay plan* (③)

There are several different types of partial pay plans, but they all revolve around the same concept. The game itself is free to access, and users are able to purchase additional items or premium subscriptions if they so wish. In some games the available items are simply aesthetic and add nothing to the gameplay beyond visual effects (such as a different costume for a character), while in other games items can give a character all sorts of special effects from faster experience gain to bonus in-game currency to unique and powerful pieces of equipment. The partial pay plan was first introduced by casual online games, but it has spread to be widely used by many types of online games.

Most Partial Pay plans entail users purchasing in-game currency using real-life money, and then spending the in-game currency on the items they desire. Some games even allow players to transfer this in-game currency amongst themselves, adding a further layer to the

Figure 2.1 One example of a Partial Pay plan (Nexon's Kart Rider)

Source: <http://kart.nexon.net/>

game's economy. Payment can be made via various methods including a mobile phone small-amount billing system, credit card, wire transfer. Many games even offer gift certificates.

4) *Set-Amount plan for Internet cafés* (④)

The set-amount plans available for Internet cafés are much the same as those offered to individual users, only on a larger scale. Under a set-amount plan an Internet café pays a monthly fee per computer for unlimited access to the game, and then charges its customers to use the computers. These plans are most economical for very popular games such as Lineage, where an Internet café might see all of its subscribed computers used for 10 hours a day or more to play the game.

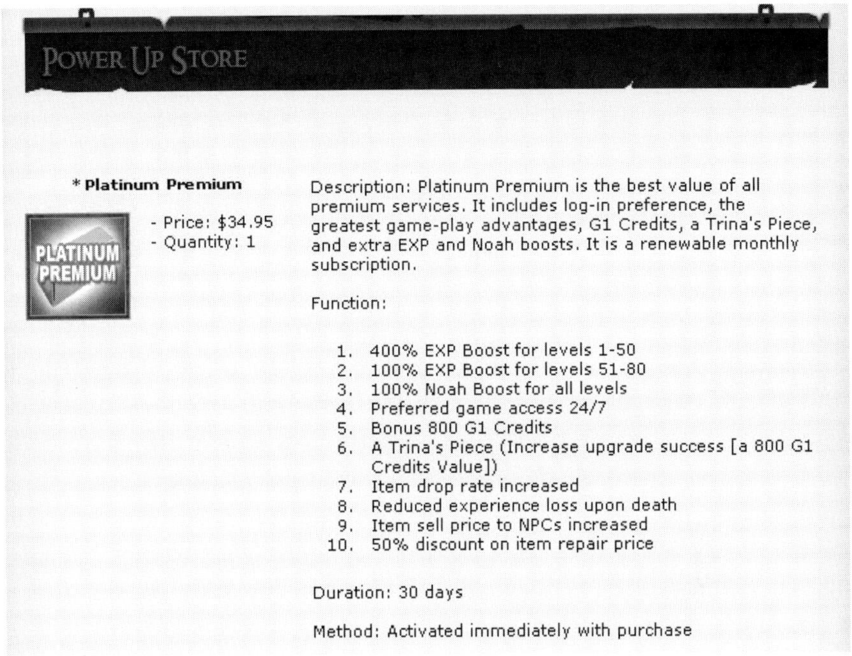

Figure 2.2 A different type of Partial Pay plan (MGame's Knight Online)

Source: <http://knightonlineworld.gamersfirst.com/powerupstore.php#MagicBag>

Table 2.4 NCsoft's Prepaid Set-Amount plan

Number of computers	Price per computer per month
1–5 computers	$66/computer
6–9 computers	$60/computer
11–19 computers	$55/computer
20–29 computers	$50/computer
30 or more computers	$44/computer

Source: <http://pcbang.ncsoft.co.kr/PayInfo/PayInfo.aspx>

5) *Set-Volume plan for Internet cafés* (⑤)

A set-volume plan for an Internet café allows the business to purchase a large block of hours of gameplay which are shared amongst all of its

Table 2.5 NCsoft's prepaid Set-Volume plan

Product	Price
300 hours	$88
600 hours	$165
1,000 hours	$275
3,000 hours	$770

Source: <http://pcbang.ncsoft.co.kr/PayInfo/PayInfo.aspx>

computers. The advantage of this plan is the ability of up to 70 PCs (in the case of Lineage) to be simultaneously connected to the game. The owner of the café also knows exactly how much they have paid for the access time, which makes determining a pricing scheme simpler. Finally, the prepaid hours do not expire, so for less popular games this can be a convenient payment system. An Internet café can use set-volume plans for several different smaller titles, which allows it to offer a wide variety of games to its users without great expense.

There are many different types of set-volume plans offered to Internet cafés by different game developers. The size of the online gaming industry in Korea means that Internet cafés have a large number of games to choose from, and game developers try to come up with enticing plans to draw in their business.

(1) *World of Warcraft*

World of Warcraft's set-volume pricing plan can be viewed as fairly standard. An Internet café purchases a block of hours which can be used on any of their computers, with a discount for purchasing larger numbers of hours at a time.

(2) *NCsoft (Lineage/Lineage 2/Guild Wars)*

NCsoft's set-volume plan leverages the company's multiple successful titles to present an enticing package to Internet café owners. The plan's hours apply to any of NCsoft's titles; rather than having to purchase plans for each game separately they are combined into one

Table 2.6 Prepaid Set-Volume price plan for World of Warcraft

Product	Price	Price per hour
400 hours	$102	$0.25
800 hours	$192	$0.24
1,600 hours	$380	$0.23
3,000 hours	$670	$0.22

Source: <http://www.worldofwarcraft.co.kr/igr/info/index.jsp>

Table 2.7 NCsoft's Set-Volume plan

Product	Price	Actual available hours
300 hours	$88	325 hours
600 hours	$165	650 hours
1,000 hours	$275	1,085 hours
3,000 hours	$770	3,255 hours

Source: <http://pcbang.ncsoft.co.kr/PayInfo/PayInfo.aspx>

package, which is extremely convenient for business owners. NCsoft also offers a 50% discount from 2 a.m to 10 a.m, meaning that only half of the actual time played during those hours is deducted from the Internet café's balance. NCsoft's plans allow up to 70 different IP addresses to connect from one location.

NCsoft also offers a separate payment plan for Guild Wars by itself. Under this plan, an Internet café can purchase a permanent license which allows unlimited access to the game. Expansions to Guild Wars are released regularly and must be purchased separately. This pricing plan can be very lucrative if a café has a large base of Guild Wars players, and there are some Internet cafés which offer Guild Wars exclusively and make use of this plan.

(3) *Nexon (including Kart Rider, Maple Story and Mabinogi)*

Nexon offers both set-amount and set-volume plans to Internet cafés. As another developer with multiple titles available, Nexon offers both

Table 2.8 Combined Set-Volume plan

Product	Fee per hour (VAT not included)	Total fee (VAT included)	IP used	Eligible games
300 hours	$0.29	$75	ALL	All Nexon games
600 hours	$0.23	$148	ALL	All Nexon games
1,000 hours	$0.22	$240	ALL	All Nexon games
2,000 hours	$0.21	$459	ALL	All Nexon games
4,000 hours	$0.20	$873	ALL	All Nexon games

Source: <http://pcbang.nexon.co.kr/>

Table 2.9 Single Set-Volume plan

Product	Fee per hour (VAT not included)	Total fee (VAT included)	IP used	Eligible games
300 hours	$0.21	$68	ALL	Any single Nexon title
600 hours	$0.20	$134	ALL	Any single Nexon title
1,000 hours	$0.20	$219	ALL	Any single Nexon title
2,000 hours	$0.19	$417	ALL	Any single Nexon title
4,000 hours	$0.18	$791	ALL	Any single Nexon title

Source: <http://pcbang.nexon.co.kr/>

Table 2.10 Premium Set-Amount plan

Product	Fee charged (VAT included)	IPs used	Eligible games
Combined set amount	$45/month	1	All games serviced by Nexon
Single set amount	$40/month	1	Any single Nexon title

Source: <http://pcbang.nexon.co.kr/>

a combined plan that allows access to all of its games, and a single-title plan for one game only. Nexon also does not put a limit on the number of eligible IP addresses, so any computer in the café can be used to play Nexon's games under their plans.

Nexon also offers two different types of set-amount plans. Similar to their set-volume plans, there is a combined plan which allows access to any of their games, and a single plan for access to one title. The fees on the Table 2.10 are monthly rates per computer.

6) *Combined price plans*

Internet cafés make natural outlets to market games, and some game developers try to encourage them to do so via profit-sharing plans. Pimang, a game portal website offered by Neowiz, offers a plan called "Pimang Internet Café Plus". An Internet Café can register for this plan and then make use of Pimang's "Say Cash Internet Café Charge" system.

Under this plan, an Internet café can purchase in-game money from Pimang in bulk. When players want to purchase items in one of Pimang's games, instead of paying Pimang directly, they buy the money directly from the Internet café. The café owner receives 10% of the profit for acting as the intermediary in the transaction, which encourages Internet cafés to promote Pimang's games. Also, when a player plays a Pimang title at a designated Pimang Internet Café Plus venue, they receive in-game benefits such as bonus experience points.

Silk Road is another game making use of this pricing plan. When a venue registers as a "Silk Road Internet café", the game company shares 10% of all profits generated at that particular venue with the Internet café. These combined plans are an interesting way for newer games to attract users, as Internet cafés can make more money if its users are playing a game in which the café shares the profits.

2. Pricing plans of various online games

Few online games are being serviced based on just one revenue model. Almost online games are using combined various revenue models. Price plans shown in Table 2.11 are the samples of online games.

1) *ROSE Online*

ROSE Online makes use of every price plan discussed, to offer the greatest number of options to individual customers and Internet café

Table 2.11 Price Plans used in Korea

Game	Price plan	Other games sharing this price plan
1) ROSE Online	① + ② + ③ + ④ + ⑤	—
2) Lineage, Lineage 2	① + ② + ④ + ⑤	Mu, RF Online, Ragnarok
3) Mabinogi	① + + ③ + ④ + ⑤	—
4) World of Warcraft	① + ② + ④	—
5) Kart Rider	③ + ④ + ⑤	Knight Online, Droyan
6) Freestyle	③ + ④	Yulgang, Fortress 2 Blue, Special Force, Gunster
7) Goonzu	③	Colossus, Maple Story, City Racer, Red Gem

Table 2.12 ROSE online individual Set-Amount plans — platinum and premium

Classification		Platinum plan	
Type	1 month (30 days)	2 months (60 days)	3 months (90 days)
Fee	$15	$29	$41
Row	300 points	600 points	1,200 points
		Premium plan	
Type	1 month (30 days)	2 months (60 days)	3 months (90 days)
Fee	$12	$23	$32
Row	100 points	300 points	500 points

Source: <http://payment.roseonline.co.kr/>

Table 2.13 ROSE online's individual Set-Volume plan

Classification	Set-Volume	
Type	5 hours	20 hours
Fee	$2.80	$7.40

Source: <http://payment.roseonline.co.kr/>

owners alike. ROSE's individual pricing plans are further divided into "Platinum" and "Premium" plans.

ROSE Online's individual set-volume plan is shown in Table 2.13.

Of note in ROSE Online's plans are the "Row points" awarded for the set-amount plans. Row points are used in-game to purchase special items, and users who subscribe for the Platinum plan pay a slightly increased monthly fee in return for more Row points. Players also earn 0.05 Row points per minute spent playing ROSE Online at an Internet café. Row points can be seen as a modified version of the partial-pay item sales model. ROSE Online also offers two price plans for Internet cafés, shown below. ROSE Online's Prepaid Set-Amount Plan and Prepaid Set-Volume Plan for Internet Cafés.

Table 2.14 Prepaid Set-Amount plan

Type	Fee per IP (tax included)	Discount	Number of IPs provided free	Payment method	Bonus Row points awarded to players
1–4 Ips	$22	0%	None	Direct deposit,	0.05 points for
5–10 Ips	$21	5%	1	Credit card,	1 minute
11–20 IPs	$19	7%	2	Wire transfer	
21–30 IPs	$17.50	10%	3		
31 IPs or more	$15	13%	4		

Source: <http://playment.roseonline.co.kr/>

Table 2.15 Prepaid Set-Volume plan

Type	Fee per IP (tax included)	Discount (%)	Payment method	Bonus Row points awarded to players
300 hours	$50	0	Direct deposit,	0.05 points for
600 hours	$101	0	Credit card,	1 minute
1,000 hours	$155	8	Wire transfer	
2,000 hours	$297	12		

Source: <http://playment.roseonline.co.kr/>

ROSE Online offers volume discounts for both types of plan, as shown in the tables. There is also another special offer for Internet cafés that make use of the set-amount plan, which is an extra number of free IPs, depending on the plan selected. For example, if an Internet café purchases a set-amount plan for 10 computers to be able to access ROSE Online, they pay a fee of $210 per month and receive one free account, so 11 computers at the café can be used to play the game.

2) *Lineage, Lineage 2*

NCsoft's Lineage games offer all payment options with the exception of an item mall. Users pay for access to the game, but are not able to purchase extra in-game items for cash. The pricing plans for both Lineage titles are displayed in Table 2.16.

Lineage 2 also offers prepaid game cards, which can be purchased from Internet cafés or other retail outlets, such as electronics stores. The card is known as the L2 Card, and comes in two different versions: the L2 7-Day card for existing players, and the L2 10-Day Coupon for new players. The L2 10-Day Coupon works much like a trial account for other online games, allowing up to 10 days of unlimited gameplay, after which a normal subscription method must be used if the user wishes to continue playing.

The L2 7-Day Card sells for $10 and functions exactly like a set-amount subscription, allowing unlimited gameplay for 7 days. NCsoft also sells a package of 2 L2 7-Day cards for $20 which comes with an L2 10-Day Coupon for free. This marketing method allows existing players to purchase gametime while also receiving a coupon to give to a friend and get them to try the game.

3) *Mabinogi (Nexon)*

Mabinogi's price plan (Table 2.17) consists of an individual set-amount plan, item sales, and both set-amount and set-volume plans for Internet cafés. The pricing system for individual Mabinogi players is quite complex, offering players a wide variety of options to customize their gameplay.

Table 2.16 Lineage 2's individual Set-Amount and Set-Volume plan

				Payment method				
Classification	Type	Fee	Discount (%)	Direct deposit	Credit card	Internet banking	Mobile phone	ARS
Individual	30 days	$29		O	O	O	O	O
Set-Amount	90 days	$70	21	O	O	O	—	—
Individual	3 days	$3	—	—	O	—	O	O
Set-Volume	5 hours	$5	—	—	O	—	O	O
	10 hours	$8	20	—	O	—	O	O
	30 hours	$18	40	—	O	—	O	O

Source: <https://secure.lineage2.co.kr/payment/default.asp#goMenu1>

Table 2.17 Mabinogi's price plan

Classification	Fee	Defining attribute
Fantasy Life Club	$18.50/28 days (33% discount)	A combined plan that gives players access to all of the features in the smaller plans listed below. Also available in one-day increments. Offers the player a discount when purchasing Premium Character Cards.
Extra Storage Service	$9/4 weeks	Offers players additional storage space for their characters in the game. Can be paid in one-week increments after the initial four-week plan expires.
Nao's Support Service	$9/4 weeks	Offers players assistance from Nao, an NPC in the game. Can be paid in one-week increments after the initial four-week plan expires.
Advanced Play Service	$9/4 weeks	Offers players a character-growth service to make their characters stronger. Can be paid in one-week increments after the initial four-week plan expires.
Vacation Packages	$34/8 weeks (25% discount)	A special package only offered occasionally, the vacation package gives players a 2-month Fantasy Life Club subscription with other special benefits such as a free Premium Character Card and a random Animal Character Card.

Source: Compiled from <http://www.mabinogi.com/>

Mabinogi offers single small plans for players who are only interested in a particular premium service, as well as a combined plan that gives access to all premium services for a lower price. Mabinogi also has a partial pay system by which players can purchase individual items, which are divided into two types: items related to character creation and reincarnation, and items related to a character's animal companion.

Character-related items allow players to create a new character or reincarnate an existing character into a new form. The two types of cards available are Basic and Premium Character Cards. Both cards allow for the creation and reincarnation of characters, however,

Premium cards offer a much wider variety of options for the character's appearance. Subscribers to the Fantasy Life Club plan can purchase Premium cards at a discount, allowing them to get Premium cards for the price of Basic cards. These cards are one-time use items, and thus create an additional source of recurring revenue.

Characters in Mabinogi have animal companions which follow them around and assist in gameplay. Animal character cards are used to create or reincarnate these animals. Each card can create a different animal character, and each animal possesses different abilities. The price of cards varies depending on the animal.

Nexon also offers set-amount and set-volume pricing plans to Internet cafés, at the rates previously discussed in this chapter.

4) *World of Warcraft (WoW)*

World of Warcraft offers individual users the choice of set-amount and set-volume pricing plans, although Internet cafés can only choose from a variety of set-volume plans. WoW does not offer a set-amount plan to Internet cafés at present.

Table 2.18 Individual Set-Amount plan

Product	Fee
7 days (1 week)	$9
30 days (1 month)	$25
90 days (3 months)	$59

Source: Compiled from <http://www.worldofwarcraft.co.kr/main.html>

Table 2.19 Individual Set-Volume plan

Product	Fee
5 hours (300 minutes)	$3.90
30 hours (1,800 minutes)	$14.90

Source: Compiled from <http://www.worldofwarcraft.co.kr/main.html>

Table 2.20 Set-Volume plan for Internet cafés

Product	Fee per hour	Regular fee (tax not included)	Total fee (tax included)
400 hours	$0.23	$93	$102
800 hours	$0.22	$175	$192
1,600 hours	$0.21	$345	$379
3,000 hours	$0.20	$610	$671

Source: Compiled from <http://www.worldofwarcraft.co.kr/main.html>

5) *Kart Rider*

Kart Rider, another online game from Nexon, is completely free for individual users. Kart Rider charges Internet cafés for access to its servers, and has an item mall that players can use. A wide variety of items are available in the mall, from aesthetic items such as custom license plates to items that provide tangible in-game benefits, such as better vehicles and speed boosts.

There are two types of currency in Kart Rider: cash and luccis. Luccis are in-game money, awarded to players for completing races. Cash is purchased by players for real-life money. Kart Rider's system allows players to play for free and still have access to everything in the item mall, but players can give themselves a boost by purchasing cash so that they can afford more items.

Another highlight of Kart Rider's item mall is that items are rented, not purchased outright. The price varies with the rental period, from approximately $1 for a 1-day rental to $5 for a 30-day rental. Kart Rider's set-amount and set-volume plans for Internet cafés are the same as those for all Nexon games, and can be found under the listing for Mabinogi.

6) *Freestyle*

There are no set fees to play Freestyle; all of its revenue comes via optional purchases. Freestyle provides a rich item mall in which players can purchase both permanent and consumable items. Consumable

Figure 2.3 Permanent items in Freestyle

Source: <http://www.joycity.com/mall/freestyle.asp?Kind=2&Sex=2&Type=0>

items come with a set number of uses, after which they are depleted and the player must purchase another if they wish to continue using the item.

Items can be purchased with either cash or points. Cash is game money that is bought with real-life money, and points are a different form of game currency that are earned by playing the game. Freestyle's item mall also offers a diverse array of purchase options in which players can combine cash and points to buy items, and can purchase different tiers of the same item with varying degrees of power.

Freestyle is free for Internet cafés, but they do offer a "Premium Internet café" plan. If an Internet café is registered for the premium service, players at the café earn double the usual number of points while playing, and can make use of special items. This pricing option provides a strong incentive for Internet cafés at which Freestyle is popular to sign up for the premium plan, bringing in extra revenue.

Table 2.21 Freestyle's price plans for Internet cafés

Number of computers	Monthly fee
Less than 50 computers	$99/month
More than 51 computers	$132/month

Source: <http://www.gamebill.co.kr/>

7) *Goonzu*

Goonzu advocates free gameplay for life, and charges no fees to either individual users or Internet cafés to play the game. Goonzu's revenue comes solely via its item mall, in which players use real-life currency to purchase "N Coins", and then spend the N Coins to buy game items.

Goonzu is also unique among online games in that it has a restriction on monthly purchase amounts, to limit excessive item purchases unbalancing the game as well as to help prevent hackers from exploiting the item mall. Purchases are limited to $66/month for minors and $220/month for adults. Adult players can submit a petition to lift the purchase restriction with a copy of their ID to verify age.

3. The feasibility of item sales in the U.S.

While item sales have been born in Korean games for years, they are relatively unheard of in the U.S. As the popularity of online gaming grows, companies are looking for new business models for their games aside from the monthly fee that is the current standard for online games in the U.S.

Currently, the only major business model in the U.S. aside from a monthly fee is a free game in which the users are exposed to advertisements. Many Web-based board games use this model. The item mall model has proven its success in Korea, and it is only a matter of time until it is introduced in the U.S. as well. A study was done in the U.S. from 2004 to 2005 to examine the potential market for item sales, and much can be gleaned from its results.

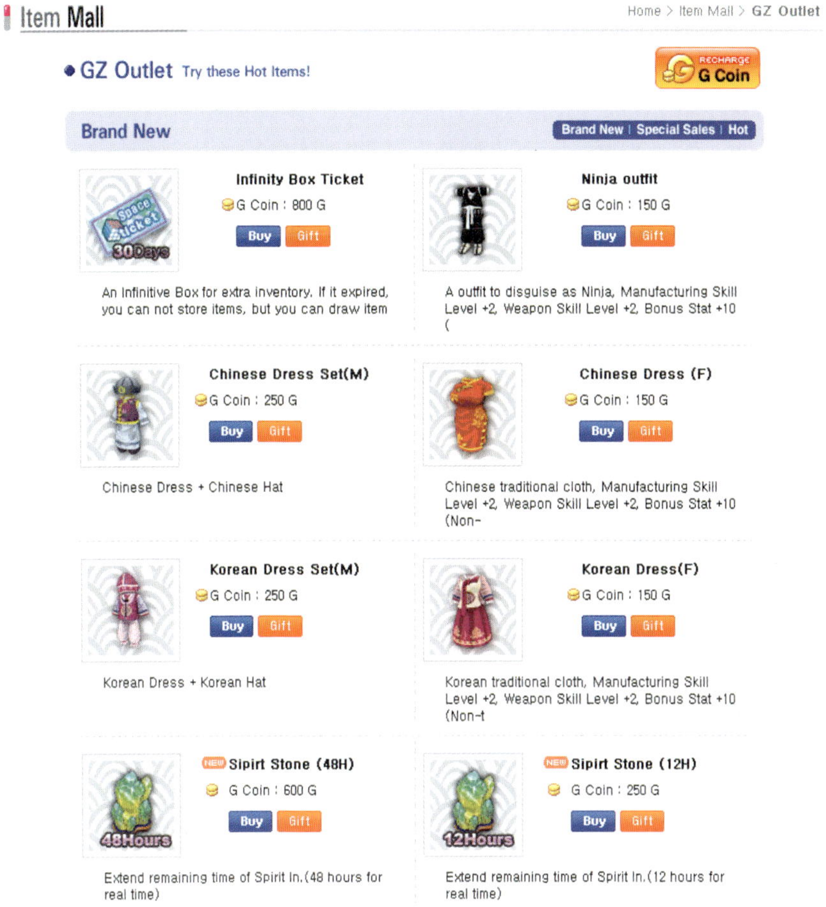

Figure 2.4 Payable items in Goonzu

Source: <http://global.goonzu.com/center/mall/product_status_list.asp?menu=86&status=N>

1) *Gender (Figure 2.5)*

When gamers were asked whether they would consider buying items in an online game, a statistically significant difference was seen in the genders of the respondents. Of the respondents who said they had no intent to purchase items in an online game, 60% were male and 40% were female. Of the group that was amenable to the idea of buying

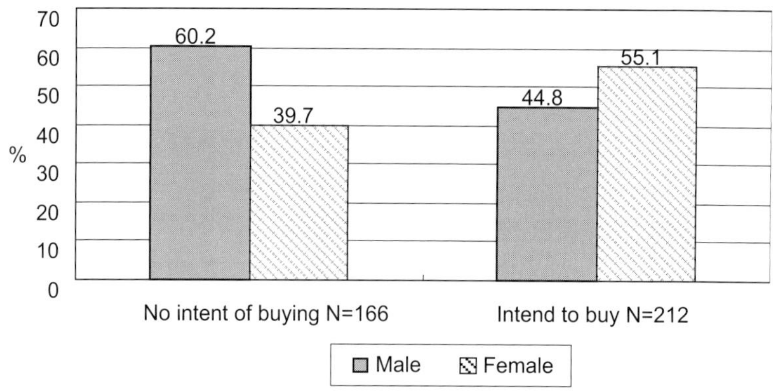

Figure 2.5 Gender

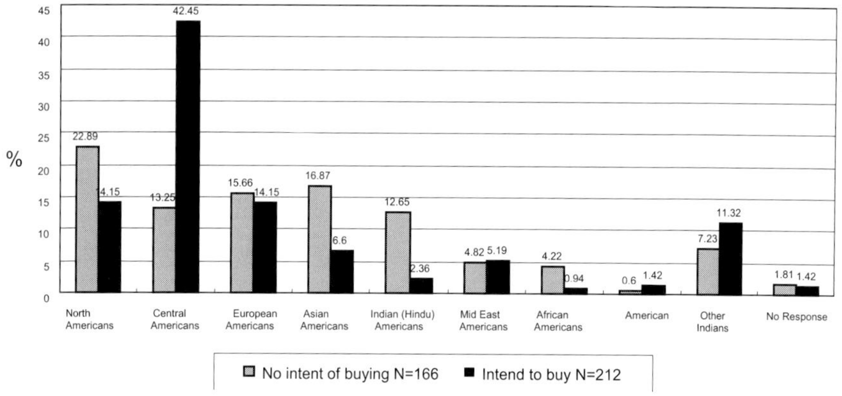

Figure 2.6 Ethnicity

items, 55% were female and 45% were male. The results suggest that marketing efforts for item sales should be directed at female players.

2) *Ethnicity (Figure 2.6)*

When the respondents of the survey were examined by ethnic group, there were clear distinctions to be found. 14.15% of both North American and European respondents said they would buy items in an online game, while a whopping 42.45% of Central American respondents said they would do so. Clearly Central American gamers are amenable to the idea of item purchases, and game developers would be wise to market to them.

3) *Type of Internet connection (Figure 2.7)*

When the survey results are examined by the type of Internet connection the respondents used, a rather glaring discrepancy arises. Roughly 20% of DSL and IDSN subscribers intend to buy items, while over 40% of respondents using dial-up modems would purchase items in an online game. While there is no definitive data to suggest why modem users are more likely to buy items, one important conclusion that can be drawn from this analysis is that users on slower connections are still a very valid target market for item purchases.

4) *Primary purpose of Internet use (Figure 2.8)*

One might expect that respondents who use the Internet primarily to play games would be the most interested in buying items in online games, but interestingly, those who use the Internet primarily to chat are even more amenable to item purchases. 55% of respondents who

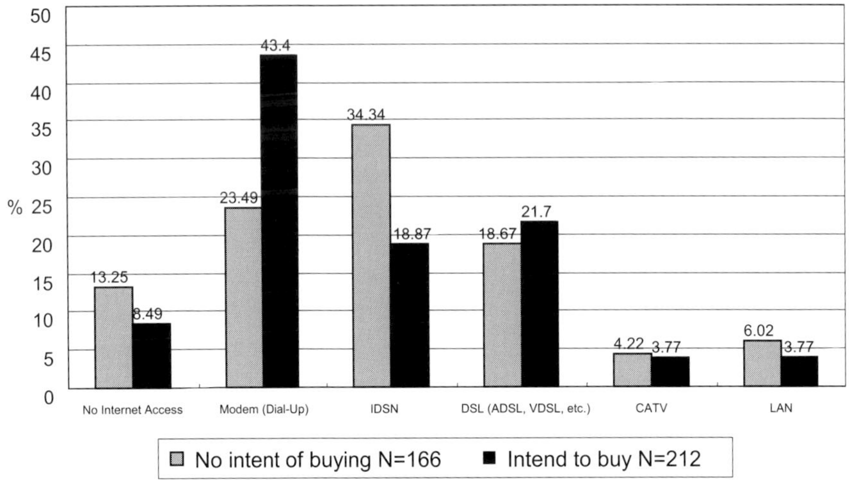

Figure 2.7 Internet connection

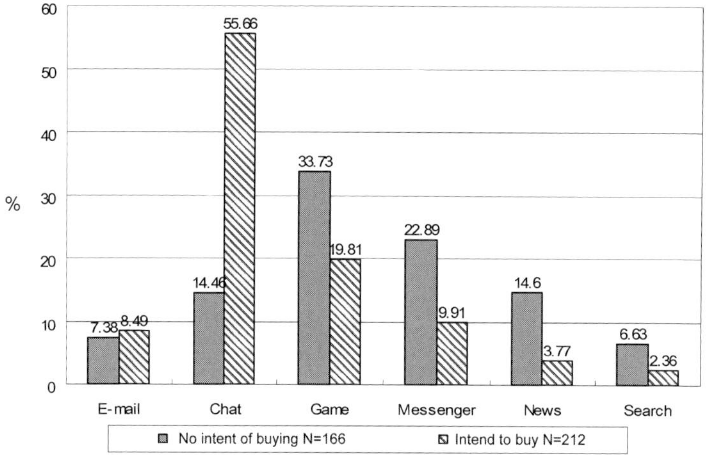

Figure 2.8 Purpose of Internet use

primarily chat online are interested in buying items, followed by 34% of gamers. This suggests that while game companies should obviously market to gamers, creating a game that draws in online chatters should also be lucrative.

5) *Preferred game genres (Figure 2.9)*

When the survey respondents are separated by which types of games they prefer, those who play puzzle games (26.42%) and casual games (23.58%) are most likely to buy items. Interestingly, only 13.21% of players who prefer card games are interested in item purchases. This stands in contrast to the Korean market, in which card games such as Go-Stop and poker attract some of the highest rates of item purchases, and shows the difference in preference in the U.S. market. Identifying these differences between the markets is key for game developers looking to implement item malls in the U.S.

6) *Game selection criteria (Figure 2.10)*

The game selection criteria analysis clearly shows that respondents who are drawn to approachable games with easy controls are by far

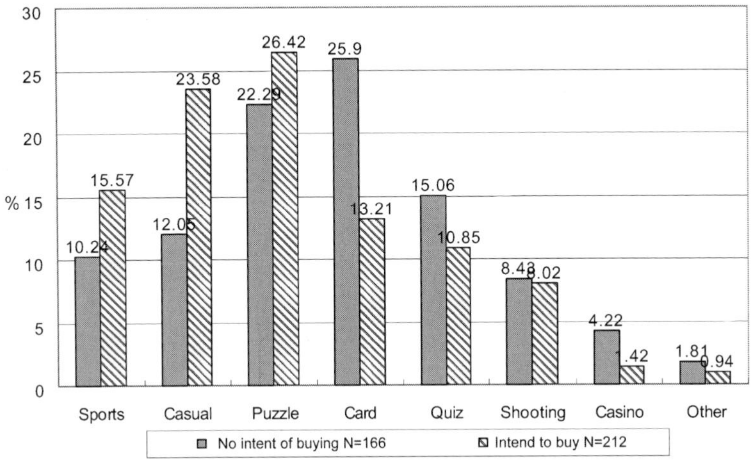

Figure 2.9 Preferred game genre

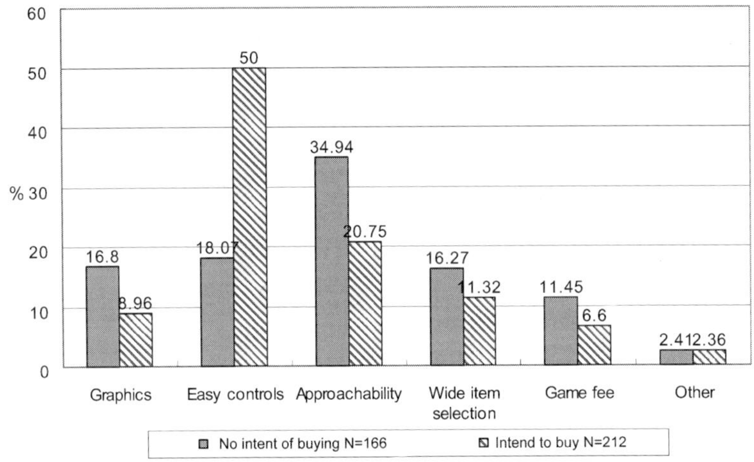

Figure 2.10 Game selection criteria

the most likely to purchase items, a finding that relates to the previously-noted inclination of casual game players to be interested in item purchases. Casual games tend to be easy to both understand and play, and these results strongly suggest that casual, simple games are good candidates for successful item-purchase business models.

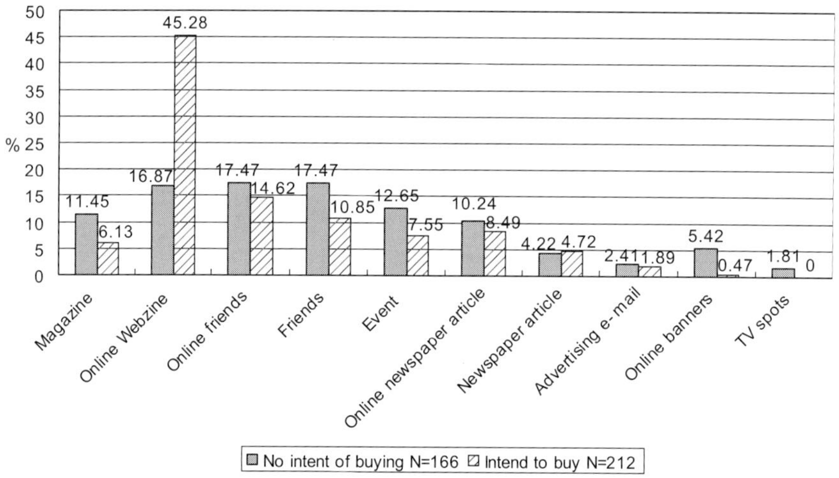

Figure 2.11 Information channels

7) *Information channels (Figure 2.11)*

This section analyzes where respondents get their information about online games. The results show that the respondents who find out about games via online Webzines and online friends are the most likely to purchase items, at 45.28% and 14.62%, respectively. These results suggest that the users most likely to buy items tend to get their game information online, and thus online marketing will be the most effective way to reach them.

8) *Primary times of Gameplay (Figure 2.12)*

Those respondents who primarily play online games in the morning, between 9:00 a.m. and 12:00 p.m., seem to be the most interested in item purchases; while those who play in the afternoon are the least inclined to buy items.

Gamers who primarily play in the morning appear to be mostly housewives or players with regular jobs who engage in a bit of light game play in the morning before work, and these groups both seem to be relatively aggressive in item purchasing.

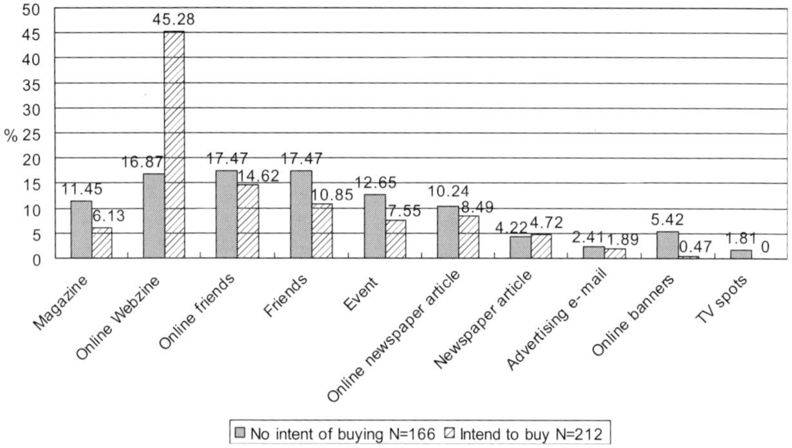

Figure 2.12 Time of day games are played

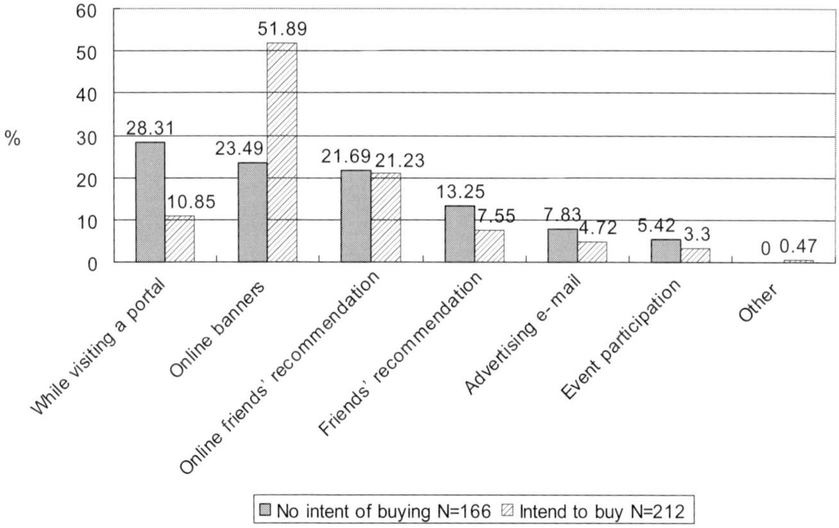

Figure 2.13 Reason for playing online games

9) *Reason for playing online games (Figure 2.13)*

This section examines the sources that cause a user to play a particular online game, and how those sources relate to the probability of item purchases. Respondents who arrive at a game via an online banner ad seem the most likely to buy items, with 51.89% of such respondents saying they would be interested in buying items. Clearly, marketing via banner advertisements should prove effective in promoting item sales.

10) *MMORPG genres played (Figure 2.14)*

Gamers who play historical online games seem like the most likely to buy items, with over 60% of such respondents saying they would make item purchases. Interestingly, two of the largest online game genres in the U.S. (fantasy and sci-fi) had the highest rates of non-interest in item purchases, both with over 20%.

There are currently only a few historically-based games available in the U.S., with the main ones being A Tale in the Desert II (eGenesis), Prince of Qin Online (Object Software), Puzzle Pirates (Three Rings Design), World War II Online (Cornered Rat), Pirates of the Burning Sea (Flying Lab Software), and World of Pirates (WOP-DevTeam). The survey results indicate that the historical game market is a rich opportunity for game developers.

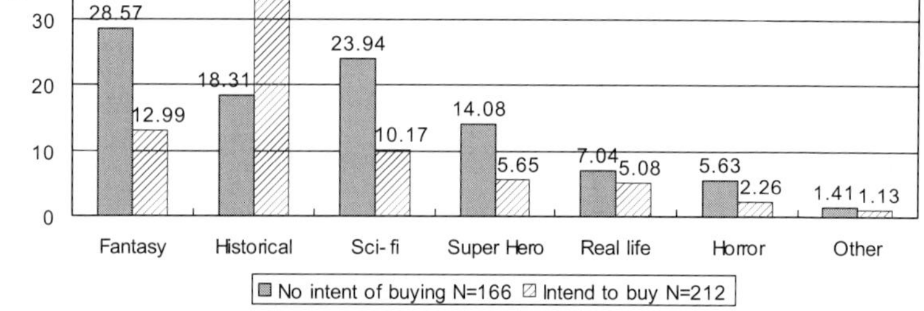

Figure 2.14 MMORPG genres played

11) *Degree of LAN party participation (Figure 2.15)*

The analysis of this section indicates that while game players who have never attended a LAN party are unlikely to buy items, gamers who are frequently at LAN parties are also not very interested in item purchases. The sweet spot for item sales seems to be gamers who have attended LAN parties, but do not do so regularly, with over 60% of such respondents being willing to buy items.

12) *Payment status (Figure 2.16)*

Respondents who play online games via an independent game developer's site are the most likely to buy items, while those who enjoy free games are (perhaps not surprisingly) the least interested in item purchases. These results suggest that gamers who already pay to play a game are the most likely to be willing to purchase items that enhance their gameplay, while players of free games wish their games to remain completely free of charge.

The most important factors to note are that selling items does seem to be a viable business model in the U.S., but that the market

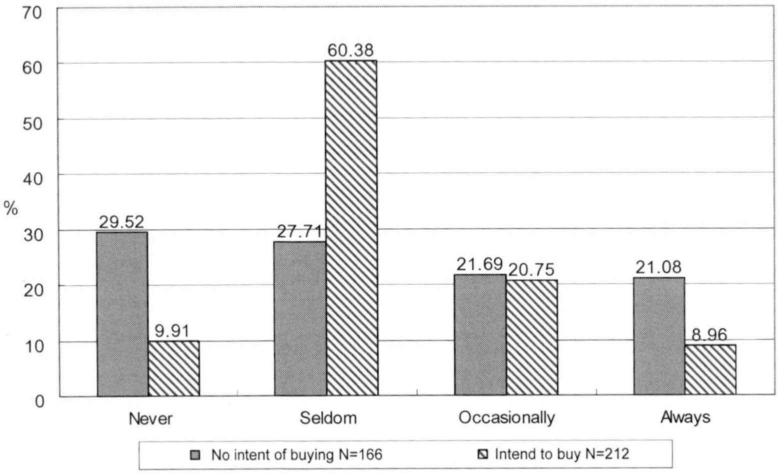

Figure 2.15 Degree of LAN party participation

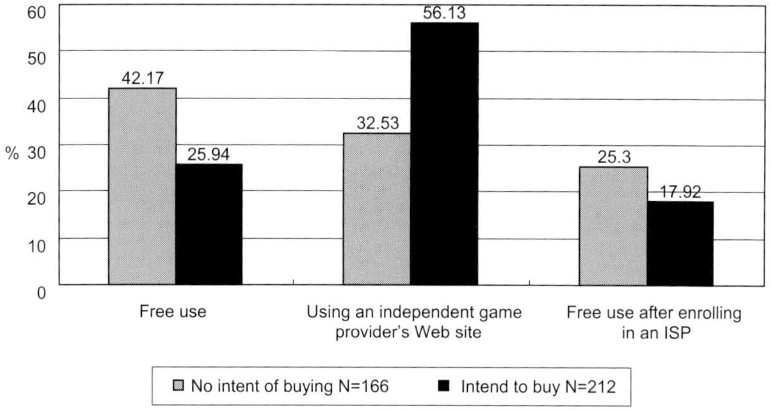

Figure 2.16 Payment status

preferences are clearly different from those in Asia. Game developers in the U.S. looking to expand their business models into item sales would do well to consider these results when creating their partial-pay plans.

Chapter 3

Virtual Societies and Economies of Online Games

Online games have evolved from a mere form of entertainment (Jenkins, 2002) into complete virtual worlds. These worlds have their own societies and structures (Castronova, 2006; 2007), and they have evolved their own economies as well, with transactions for digital assets taking place within them. These economies are growing rapidly as players find it increasingly worthwhile to spend real-life money on in-game items. The Korean item trading market, for example, is predicted to reach $1.2 billion in 2009, eclipsing the $1 billion market size in the previous year.

Online games are still a form of entertainment, similar to console and PC-based offline games (Huizinga, 1955). They are also more than that, though, as they foster virtual societies and economies. The term "virtual world" is an apt term, as online games possess many of the traits of the real world, simply in a virtual aspect.

The online game industry has provided a marketplace for online economic activities, and has created a complete virtual economy that runs in tandem with the real-world economy (Wi, 2003; 2006a). A given online game's economy could be viewed much like a foreign country, with an exchange rate between the local real-world currency and the game's currency. Cyber economies are growing more linked with real-world economies, and this growth shows that online games possess the potential to create a new industrial paradigm.

1. Generation of online economic activity

The community aspect of online games is the key factor driving their economic growth. Characters can acquire items and experience in offline RPGs as well, but these items are limited to their one character, and cannot be shared with other players. As an offline RPG involves only one player, the value of items earned in such games is extremely limited.

This valuation is completely different in an online game world. In an online game, players gain experience and items for their characters via interactions with other players, cooperating to kill monsters and collect the items they drop. As discussed earlier, items serve two primary functions in an online game: to strengthen the player's character, and for bragging rights with other players (Wi and Nojima, 2003). Players will commonly find items that their own character cannot use, and wanting to trade or sell those items to other players so that they can then purchase what they want. This creates economic transactions, and it was only a matter of time until these transactions were not bounded with the game world, but expanded to include transactions for real-world money.

Marx explains surplus value and exchange in the real economy, the exchange path induced by currency, and the process of the birth of currency below:

> "The first stage in an object to be used turning into latent exchange value is the object to be used being rendered valueless, in other words, existing as the use value in a volume exceeding that possessor's direct desire... The exchange ratio in volume of these objects was completely coincidental at the beginning. In the process the other's desire for the object to be used gradually strengthens. The endless repetition of exchange turns exchange into a normal social process... The necessity of this form improves as the number and diversity of commodities entering the exchange increases. In trades, where possessors of commodities exchange and compare their commodities with various other materials, various commodities of various commodity possessors are not exchanged

or compared to the value of a third kind of commodity. This third commodity is currency."[1]

Marx's explanation of how the real-world economy worked can also be applied to transactions in online games. Below, "objects of use", "objects", and "commodities" are replaced with "items."

> "The first stage in an item turning into latent exchange value is the item being rendered valueless, in other words, existing as the use value in a volume exceeding that possessor's direct desire... The exchange ratio in volume of these items was completely coincidental at the beginning. In the process the other's desire for the items gradually strengthens. The endless repetition of (item) exchange turns exchange into a normal social process... The necessity of this form improves as the number and diversity of items entering the exchange increases. In trades, where possessors of items exchanging and comparing their items with various other materials, various items of various item possessors are not exchanged or compared to the value of a third kind of item. This third commodity is cyber money."

The theory fits an online game's economy rather neatly. The process of item acquisition, exchange, and transactions in an online game fit the real-world model of commodity transactions (Wi, 2007a). At an economy's inception, a simple barter system of item trading takes place, and then as the economy grows more sophisticated, an exchange system based on the game's currency emerges. The final stage of growth when virtual items are bought using real-world money. The development of online games has fostered these virtual economies, and created a new economic niche.

2. Development and evolution of the item trade industry

The item trade industry has emerged only recently, but has undergone rapid growth. In 2004, the Korean item trade market was approximately

[1] Marx (1992), p. 112.

$800 million. Of this, $400 million represented revenue earned by item brokers, and the other $400 million was from transactions between individual players. The market is expected to continue growing quickly, surprising $1.5 billion by 2008.

ItemBay, one of the largest item broker, has projected transaction volume of $190 million in 2003, $270 million in 2004, and $350 million in 2005. ItemBay acts as the middleman in transactions between players, ensuring security and taking a cut off each transaction. These numbers are of interest because they surpass the annual revenue of NCsoft, the largest Korean online game developer, which reported sales of $250 million in 2004. Virtual economies are expanding at a torrid pace, and it seems that many players are very willing to pay real-world money to further their character's progress in online games.

Figure 3.1 illustrates the current state of the Korean item trade market. Item Mania boasts a 47% market share, followed by Item Bay with 45% and Item Play Forum with 4.6%.

Item brokers have a relatively simple business model. A seller lists an item for sale along with the price. If a buyer agrees to purchase the item, their account with the broker is debited for the amount of the purchase, and held in escrow by the item broker. The buyer and seller are issued an authorization code, and the seller gives the item to the buyer within the game world. When the buyer confirms that they

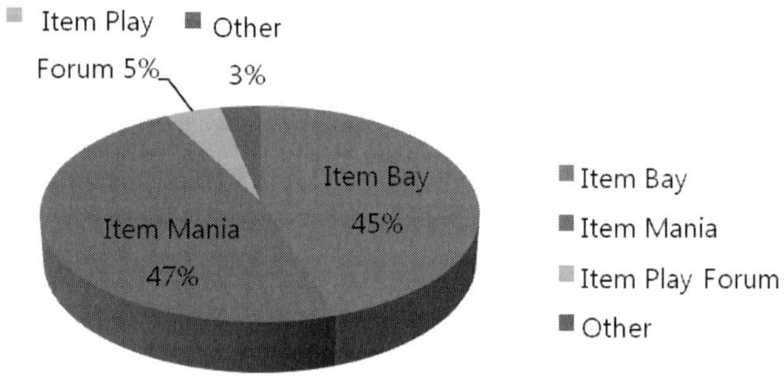

Figure 3.1 Current state of the domestic item trade market

Figure 3.2 ItemBay buy/sell postings

have received their item, the item broker releases their escrow, and the seller receives the money. For acting as the middleman and securing the transaction, the broker collects a commission of approximately 5%. Figure 3.3 graphically illustrates the process in detail.

The number of item transactions tends to increase as new big-name titles are released. For instance, ItemBay reported a 20% surge in transactions following the open beta of Lineage 2. With a new game release, some players plan to switch over, and others worry that the new game will cause a price drop for items in the old game. Lineage 2 was likely to draw players from the original Lineage, and thus as Lineage 2 drew closer to its commercial launch, item brokers saw a sharp increase in listings of Lineage items. This is an example of a completely free market economy with no overseeing body imposing restrictions. The shifts in prices to match supply and demand are fascinating.

Figures 3.4(a)–(d) show the fluctuations in the price of game currency for several different games. While Lineage's adena remains relatively stable, the prices for in-game money in RF Online and

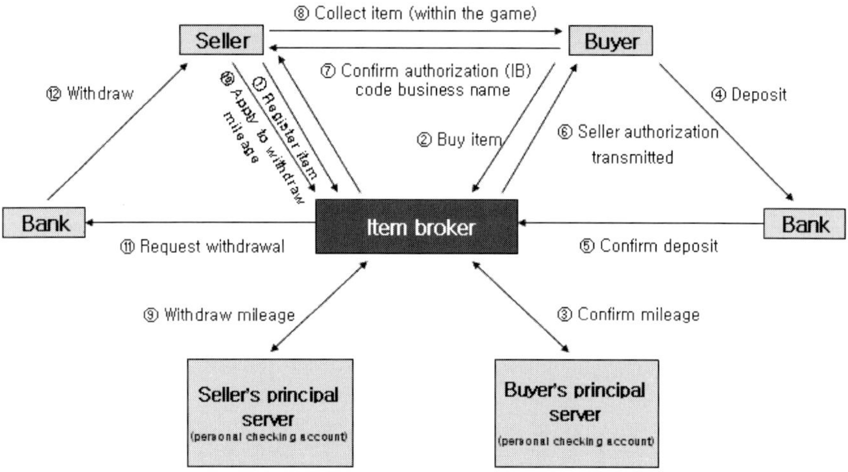

Figure 3.3 Item trade process

Legend: Numbers indicate the order in which transactions take place

Han Game are subject to sharp swings. Currency depreciation indicates an abundance of currency within the game. In other words, online game worlds can suffer from inflation just like the real world. Inflation has the potential to completely topple a game's economy if it gets out of control, so it can be helpful for developers to keep watch on the exchange rates of their game's currency to ensure that the economy is functioning properly.

Item brokers have also emerged in the U.S. and Japan. Japan's Item Bank operates on the same model as ItemBay, and its business has been growing steadily. Currently, cash transactions account for far more of the market than item trades do. As the online game market in Japan continues to expand, there will likely be a robust market for facilitating item trades as well.

The item market has appeared in the U.S. online gaming industry as well, primarily in the MMORPG genre. Ultima Online, for example, has real estate auctions for prime pieces of property within the game world. However, many game developers are wary of enabling transactions for real-world money, due to both game balance issues

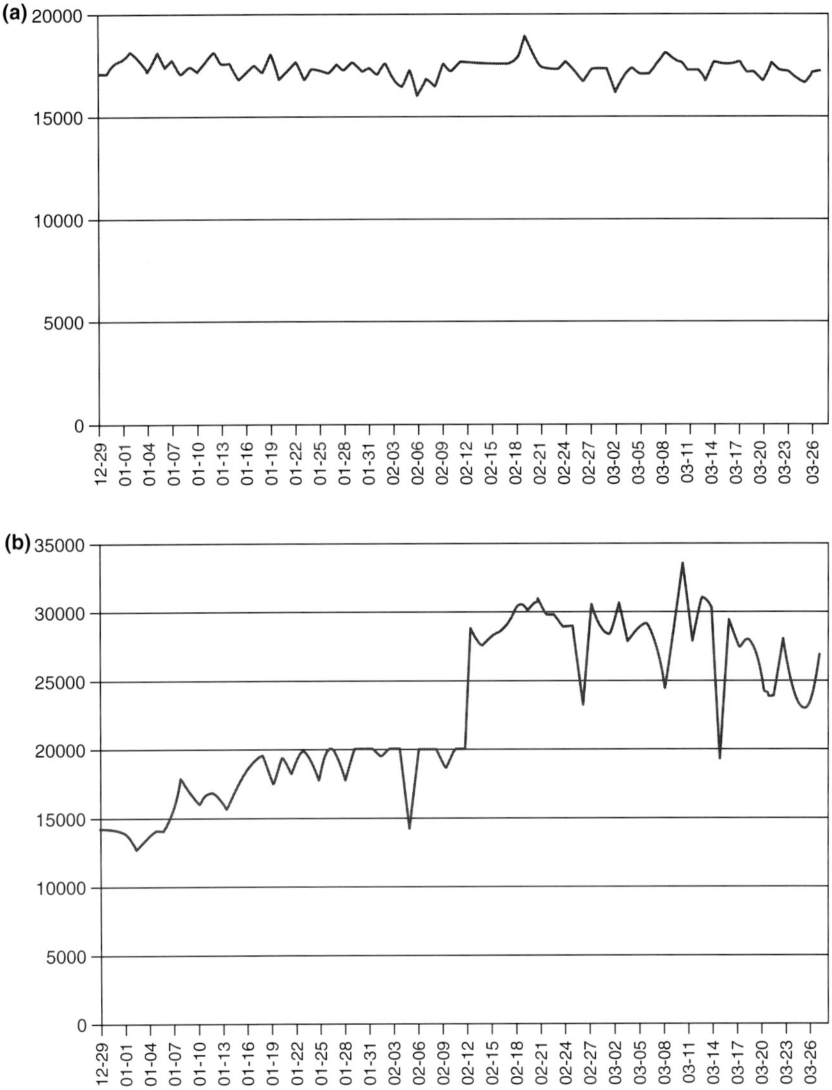

Figure 3.4 Exchange rates of several major online games (a) Lineage, Guardria server (Unit: 1 million Adenas) (b) Han Game Go-Stop (Unit: 1 billion game money) (c) Maple Story, Demedos server (Unit: 10 million Maso) and (d) RF Online Solus Bella server (Unit: 100 million Dalant)

Source: <ItemBay.co.kr>

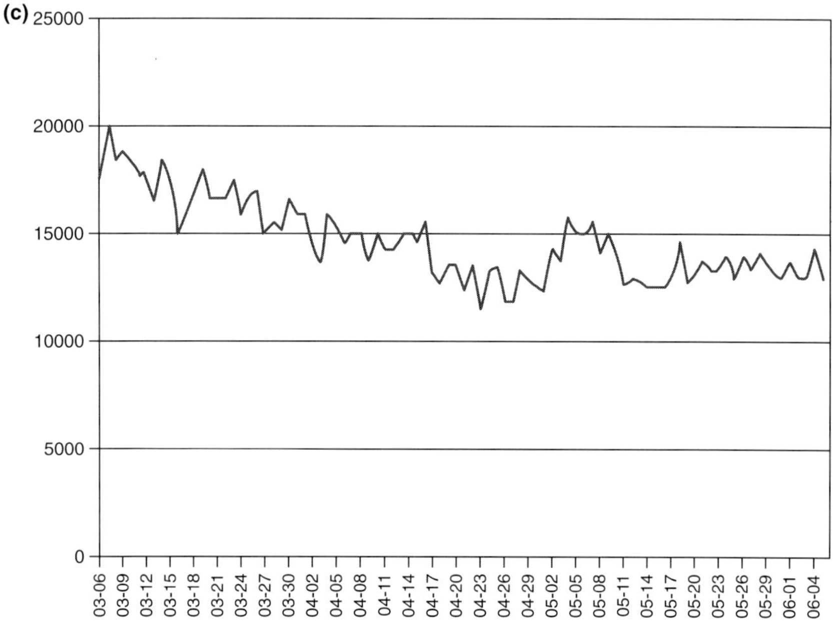

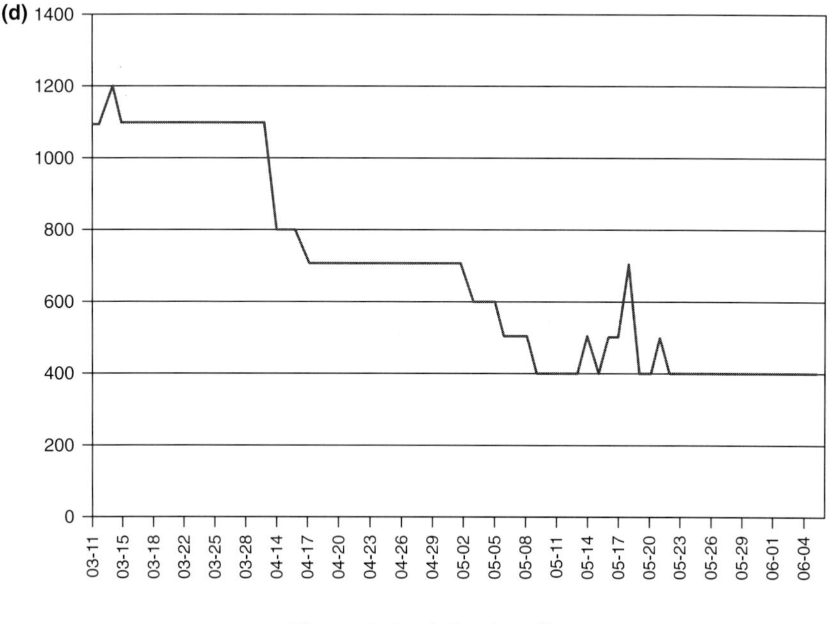

Figure 3.4 (*Continued*)

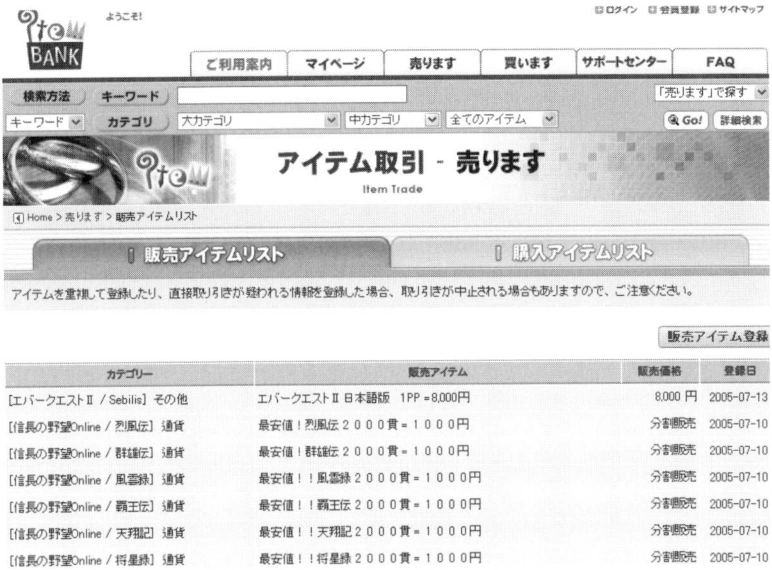

Figure 3.5 A Japanese item trade website

Source: <http://www.itembank.co.jp/>

and potential legal liabilities. Furthermore, many MMORPGs allow for mods (modifications) to be used, which are add-ons to the software that change its appearance. These mods are usually free, which limits the growth of any sort of avatar market for online games.

In the U.S. market, SOE (Sony Online Entertainment) has led the way in the item market. In July 2005, SOE introduced a new service called Station Exchange in which users on one specific Everquest II server could buy and sell items, currency, and characters. SOE does not sell virtual goods itself, but is willing to act as a broker, similar to ItemBay. With the game's developer overseeing things, users gain the advantage of the security in transactions. SOE has several regulations on this service, which are listed below:

— The service is only available on one specific trade server.
— SOE does not sell items.
— Users who wish to participate transport one or multiple characters to the trade server.

— Examining transporting characters created in the trade server to standard servers.
— Users do not possess ownership of characters, items or coins.

SOE implemented Station Exchange to capitalize on the market for item trading while avoiding the controversy about virtual item ownership by not directly selling items.

Project Entropia is another online game experimenting with virtual real estate sales. An Australian player who goes by the game name of Deathifier bought an island in the game for $26,500 at an auction, giving him full ownership of the land as well as hunting rights and excavation rights in its mines. He can now recoup his investment by selling smaller plots of land on the island to other players and collecting taxes from players who want to hunt on the island or dig for gold in the mines.

Over 200,000 users play Project Entropia, and instead of paying a monthly fee for access to the game, they buy space in which their characters live. The Project Entropia Dollar (PED) is fixed at a 10 to 1 exchange rate to the U.S. dollar, and the developer provides for exchanges between the two currencies. Project Entropia is in effect testing a new type of fee model, and is seeing success by doing so.

There is significant demand for item trades and sales in the U.S. and Japanese markets, and the potential for growth is high. However, the path and degree of growth will vary depending on the social acceptance of online games, as well as the cultural backdrop of each nation.

3. Degree of item trades and player conception of Korean users

Buying, selling, and trading items are essential parts of online gaming. Almost every player participates in the game's economy to some degree. But how many players engage in actual cash transactions? According to a study done in 2004 (as shown in Figure 3.6) 20.9% of Korean teenagers and 9% of Korean adults have purchased items for real-world money.

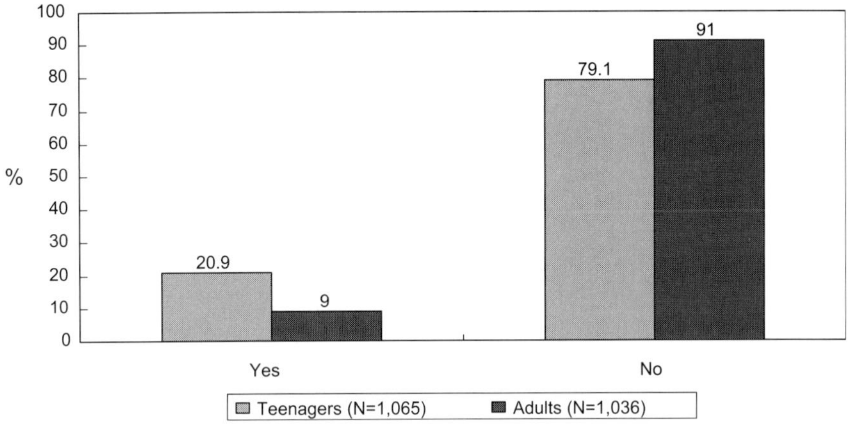

Figure 3.6 Item-Cash transaction experience

The commonly held belief in Korea is that a significant majority of teenagers purchase items in online games, which this study debunks. Even in Lineage 2, which accounts for a large proportion of the item transaction market, only one-third of users have purchased items. More than 80% of that third engage in no more than two transactions per month, spending less than $20. Though cash transactions are a large growing market for online games, it seems clear that the majority of players stay within the bounds of the in-game economy and do not spend real-world money on items.

In terms of the types of games in which users are purchasing items, teenagers primarily buy items for MMORPGs, while adults most commonly buy items for online board games. The Figure 3.7 demonstrates this.

This difference is most likely due to the kinds of games primarily played by these age groups. Teenagers most often play MMORPGs, whereas adults more commonly play board games, including online card and other gambling games.

Another important issue is how users perceive item purchases. Online game users do not just play a game in a world created by developers; their actions and perceptions have direct consequences to how that world evolves. What the players want will significantly

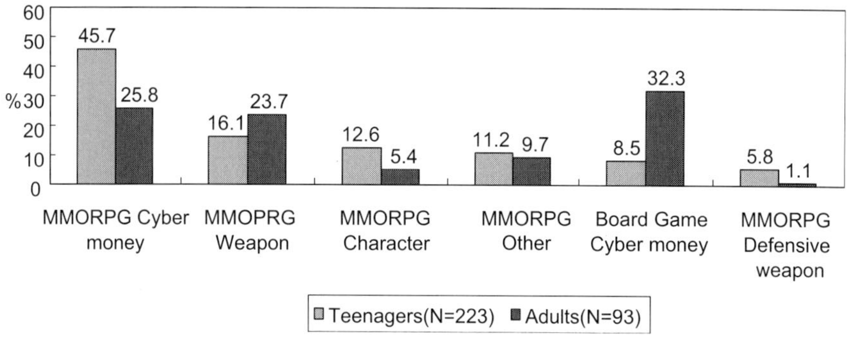

Figure 3.7 Primarily traded item category

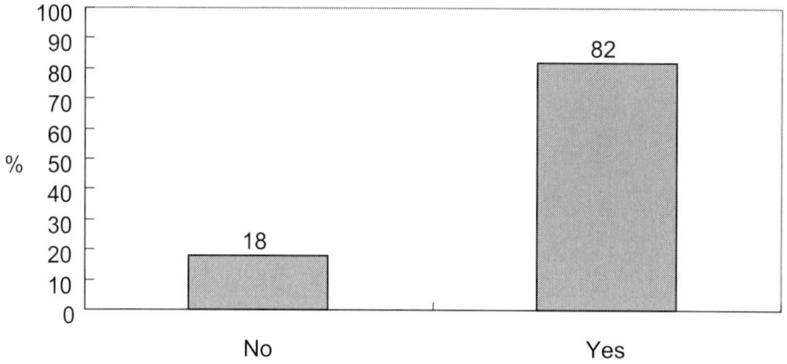

Figure 3.8 Perception of Item-Cash transactions: should be left to individual discretion

<Teenagers N = 1,065 Adults N = 1,036>

affect the direction that developers take with changes to the game. Whether the players accept or reject item purchases as a valid model in a game is extremely important. In Figure 3.8, survey results show that over 80% of users polled said that they prefer to have item transactions left to individual players, rather than being run by the developer.

An interesting result of these surveys is when users are asked about the existence of item transactions, approximately 50% of users approve of them. But, when asked about the actual method of how

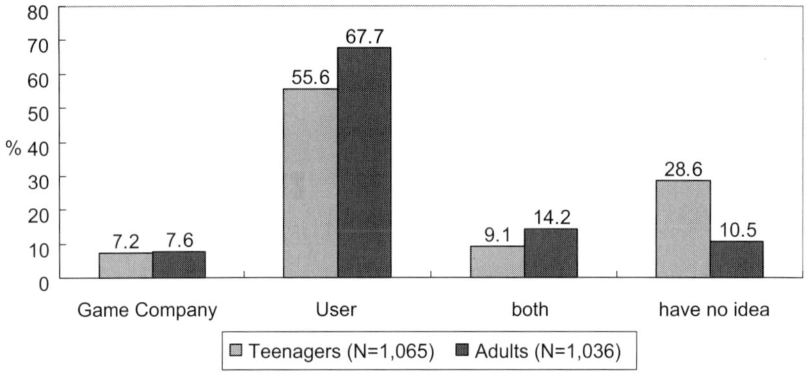

Figure 3.9 Perception of item ownership

item transactions should be handled, over 80% of users view any intervention from the game's developer with disfavor.

Another important aspect of item transactions is the concept of item ownership. Who actually owns the characters and the items? Most game developers have a term in their user agreement that states that all characters, items, and any other virtual property within their games belongs to the developers, and that players pay for access to the game and nothing more. However, 55% of teenage users and 67.7% of adults view their character's items as their own personal property. The concept of virtual item ownership has already caused some controversy, and will likely be a source of conflict between players and developers in the years to come.

The key issue for developers is whether player perceptions on item transactions are fixed or adjustable. Developers also need to know the views held towards the markets they are expanding into, so that they can implement policies and market their games accordingly. Some hints as to the perceptions of different user groups can be gleaned by comparing players from the U.S. and China. U.S. online gamers do not commonly engage in cash-item transactions, while that is very common for Chinese players.

The differences in how groups of users view item transactions is linked to whether those users value the game process or the outcome. As discussed earlier, the process takes precedence for users in the U.S.

and Japan. The actual path of completing a quest is more important to them than rapidly increasing the level of their characters. However, Chinese users tend to value the outcome, thus are more concerned with quickly leveling their characters with powerful items. Korean users tend to fall somewhere in between the trends exhibited in the U.S. and China.

There is also a difference in how these groups value the community aspect of online gaming. Users in U.S. and Japan tend to be very community-oriented, and place a lot of value on the social aspects of the game they are playing. Chinese users are generally more fixated on character levels and experience points than communities.

Item trades are clearly more than simple business transactions, and the views regarding them reflect the different cultures in different countries. Game developers would be wise to consider these differences when they are building their business models, lest they follow in the footsteps of Everquest and Ultima Online, successful U.S. games which failed miserably at expanding in Korea and China.

Many game developers have already entered the item transaction market by using the partial pay model of item sales. In the Korean and Chinese markets, charging a monthly fee for access to a game is growing increasingly difficult as more games are released and competition grows fiercer. Many game developers have looked to item sales to earn revenue, and quite a few of them rely solely on item sales and have done away with an access fee entirely. These developers are already involved in the item transaction market, and are contributing to training users to be familiar with paying real-world money for virtual items.

In the future, the economic model might well switch from the current system of user-item broker to a three-party relationship of developer-user-item broker. Such a system would provide additional security and comfort to users, who could purchase items with confidence, knowing that the developer is monitoring transactions to ensure they are completed fairly. Figure 3.10 shows statistics for several games that offer item sales, and the different categories of transactions that their users engage in.

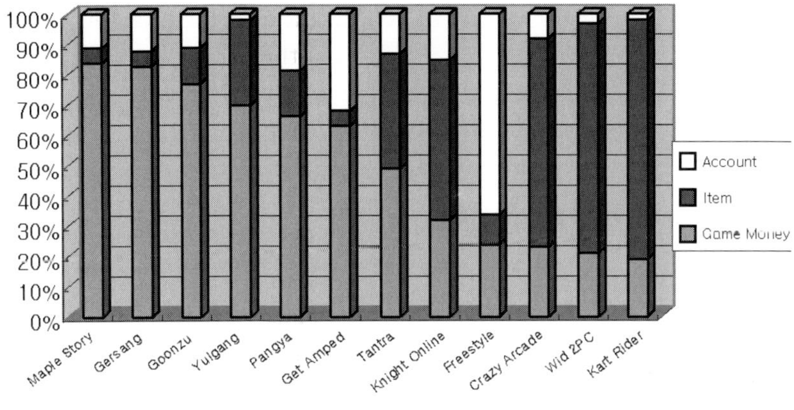

Figure 3.10 Partial Pay online game Goods Traded

Source: <ItemBay internal data>

4. Cyber economy growth model

The current item transaction model has yet to evolve into a full virtual economy, but it has transcended the simple model of users trading items between themselves and evolved into the model shown in Figure 3.11. Parts of the cyber economy and the real economy are beginning to be linked via cash transactions. Figure 3.11 shows the current item transaction model and how it links to the real economy. This system possesses several distinct attributes, as discussed in further detail below.

The first attribute is a diversity of digital content and the growth of trade in digital assets. Digital item transactions are passing beyond mere online games, evolving into a marketplace for a variety of digital assets and content such as designs, images and knowledge. To date, the market structure consisted of search portals, online games, avatars, design, and knowledge creating their own autonomous market segments, each with their own groups of users. In the future, these genre demarcations will collapse and the segments will fuse together with an evolved item transaction system as the basis. Currently, Naver Knowledge Search (NHN) is an example of this evolved digital item transaction system.

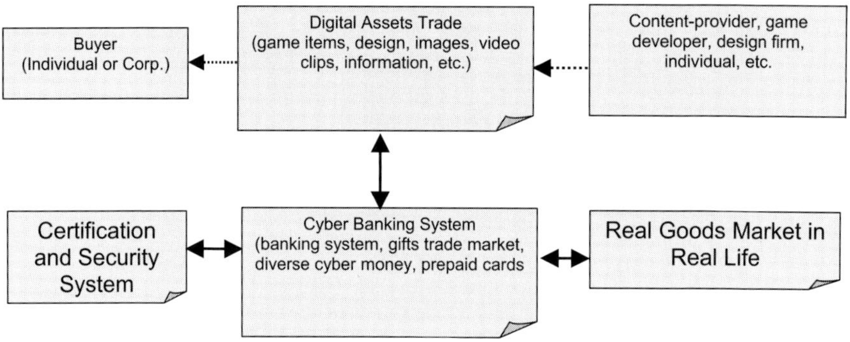

Figure 3.11 Evolved item trade system

Legend: <◄······· Flow of digital assets, ◄► module interaction>

The second factor is that online game item transactions possess the potential to develop into full cyber markets, linked to online banking. Currently, item brokers act as the middlemen in transactions and handle money digitally for their users. These brokers act much like online banks, essentially providing a savings account in which users can store their money digitally. As time goes by, item brokers may start offering interest on cash sitting in an account as well as offering other financial services, perhaps by partnering with online banks. It is also likely that other services will begin to be offered by item brokers, such as rentals or gifts.[2]

The emergence of cyber banking will also require more advanced security features and certifications. Current online game server structures are vulnerable to hacking, which discourages users from spending real-world money if they are afraid that the transaction may not be secure. Higher-quality security measures and certifications will reassure users of the safety of their transactions, and encourage more users to purchase items digitally.

Third, the growth of the item transaction market will promote the growth of other digital content providers. Companies that provide digital design services, images, and of course game developers will

[2] In late 2003 game developers decried ItemBay announcing item rentals. But when seen in terms of the cyber economy, item rentals and trades are materially identical. In other words, both methods are in essence selling elements of the game for monetary value.

profit from the growth of the item transaction market. The growth of avatar-design companies is a good example of this, with luxury avatar-item brands such as Louis Vuitton and Barbie being recently offered to online gamers.

The fourth attribute is the link between digital assets and real-world goods. Some cyber goods can already be exchanged with real goods. Some portals such as Freechal issue cyber money and then sell products that users pay for with the cyber currency they have accumulated.

Cyber currency will also likely evolve further. Currently, some forms of cyber currency issued by companies are linked together and can be exchanged. However, a "cyber key currency" that can be commonly exchanged with any real-world goods has not yet emerged. The various types of cyber currency are akin to regional currencies, which can only be used in select transactions and within certain areas. In future, however, a market to exchange all of the various types of cyber currency will likely emerge, possibly fostered by currency issuers themselves. This would allow cyber currency to take its place among the various real-world currencies of the world's nations, and be exchangeable just as they are.

5. Issues to be considered in item transactions

1) *Item ownership*

Item ownership is a common cause for disputes in digital transactions, and these disputes arise from two different angles. The first is the dispute between the game developer and the user over the degree of item ownership, and the second is the dispute arising between users.

Most online game players view their character's items as their personal property, but game developers have user agreements which declare that all in-game items are their property. This dispute has been going on for years, and will only escalate in the future as cash transactions as in-game items become more common. This dispute is also likely to cause difficulties in any legal action arising from item transactions, such as accusations of fraud or deception.

The central element in these two disputes is the "materialness" of items. Korean Article 98 of the civil code defines objects as natural forces such as material things or electricity which can be managed. Thus, material things are things with shape which take up physical space and can be detected with the five human senses. Shapeless objects include electricity and other natural forces cannot be controlled. Virtual items are thus not included as material objects under the current law. No matter how much an item may be worth in an online game, it is not considered a material good under the current civil code.

Under the criminal code, Korean Article 329 stipulates that "individuals who seize the property of others are punishable by a sentence of less than 6 years or a fine of less than $10,000." For larceny to be established in a court of law, the seizing of the personal property of another has to occur. However, since game items are not considered property, stealing items in an online game is not punishable by law. Under the current law larceny, embezzlement, stealing, and destruction, all of which require the establishment of materialness, cannot be brought to court for cases involve game items.[3]

Should game items be acknowledged as material goods and at least partial ownership of items awarded to players by the court system, game developers will face a very difficult situation. Developers will be saddled with an enormous financial burden in managing users assets, as they could be held responsible for any items lost due to a service outage, or even potentially sued if a content change to the game reduces an expensive item's value significantly. They will also lose their authority over the existence of their game, as they would not be able to shut down a server without offering compensation to any users affected.

---

[3] In August 2002 the Korean Supreme Court decided no larceny is established in a case involving an employee who extracted a secret blue print from the company lab acting on the order of a former executive. In the decision the majority opinion stated in order for larceny to be established, the stolen object must be considered property, but information stored in a computer is not considered material property, nor is it a natural force transcending materialness. The court also held as extracting or duplicating information does not compromise the information *per se*, its possession or possibility of use, no larceny is established.

2) ***Credibility in item trades***

The current system of item transactions is not based on the same economic logic we see in the real world. One factor restricting rational economic trade is the issue of credibility in item transactions. Due to the common occurrence of hacked items in online games as well as severe restrictions implemented on item trading by many game developers, the credibility of the parties involved in a transaction trumps normal economic reason. In most economic decision-making, price is the most important factor when choosing between vendors who provide the same product. When it comes to online item transactions however, credibility trumps price significantly, even when an item broker is involved.

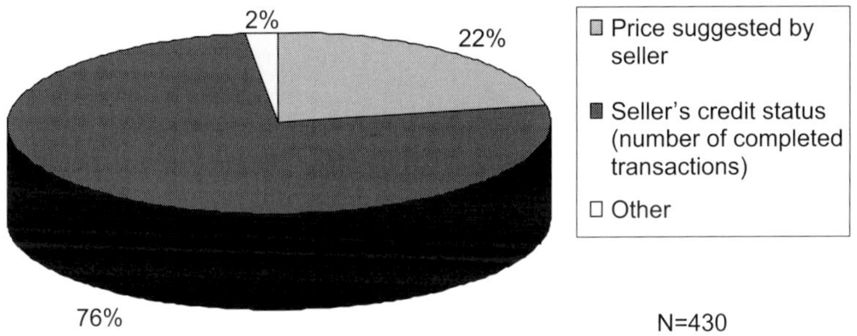

Figure 3.12 The most important element to consider in Item-Cash transactions involving item brokers
<Korean users N = 430>

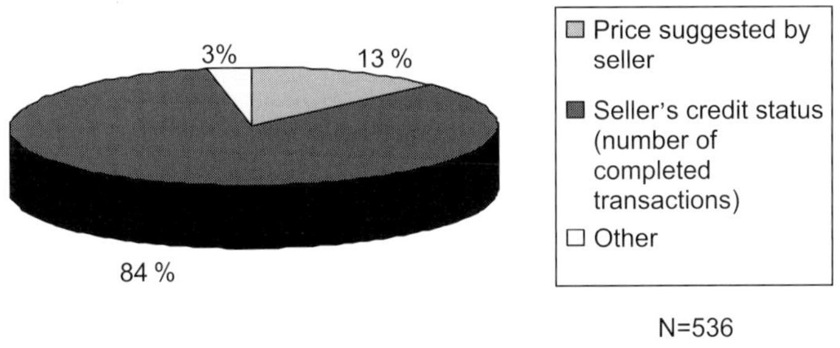

Figure 3.13 The most important element in direct Item-Cash transactions
<Korean users N = 536>

A core online gamer (Internet café owner, 5 years experience playing online games) commented on these phenomena below.

> "The central issue at hand is numerous accounts or Adenas being traded in Lineage have been hacked hence buyers must be wary of sellers' credibility, and they desire secure transactions. NCsoft also opposes item cash transactions, and when such transactions are detected, they take drastic measures as to seize accounts, and such policies are a product of item buyers being concerned about the credibility of sellers."

The credibility issues in the digital item market serve to impede the growth of the field. New traders need to spend time as buyers to build their reputation, which limits the number of trusted sellers. There are also concerns regarding professional item sellers that collect their inventory via illegitimate means such as macro use.

3) *Game system and item trade*

The way that item systems are designed in online games strongly influences the nature of item transactions that will take place in those games. ItemBay's listings for several different games are displayed in Figure 3.14, and serve to demonstrate the truth of this statement. World of Warcraft, Lineage, and Lineage 2 are all MMORPGs, but the volume of trades is dramatically different between the two Lineages and WoW. Lineage and Lineage 2 have significantly larger subscriber bases than WoW, with 130,000 and 90,000 simultaneous users compared to WoW's 40,000. However, item transactions in WoW are dwarfed by those in the two Lineages, by a degree far surpassing that which would be expected due to the population difference.

The difference lies in the item system used in World of Warcraft. WoW has a system called "soulbinding", by which items become permanent possessions of the player who uses them. The most powerful items are soulbound to a player as soon as they are picked up, though there are also other types of items that are only soulbound when a player actually equips the item. These items can be sold to another player, but, only until they are equipped for the first time.

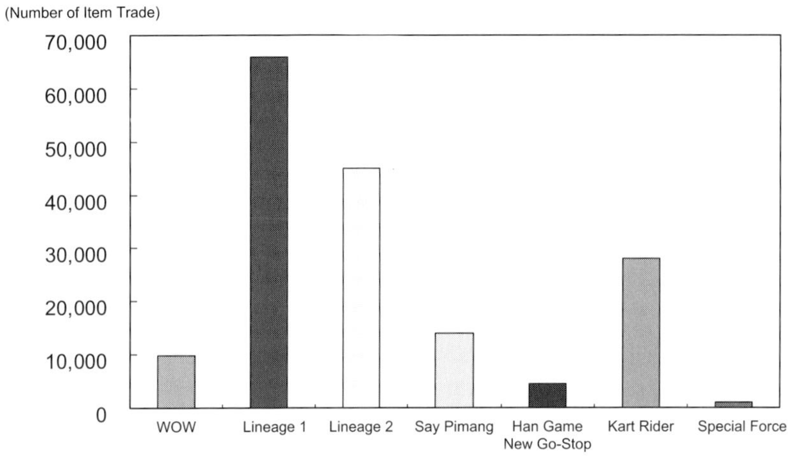

Figure 3.14 Items traded on ItemBay monthly

Source: ItemBay internal data

The item system in World of Warcraft is significantly different from other MMORPGs, in which items can be freely traded even after being equipped. WoW's developer, Blizzard Entertainment, implemented the soulbinding system specifically to prevent item transactions from taking place, and to stop players from simply buying whatever equipment they want for their characters. This item system has limited cash transactions in World of Warcraft to game money and accounts.

Other game developers who do not want to see item transactions taking place within their games could implement a similar system to the one in World of Warcraft. They should be careful however, as not allowing item transactions could cost subscribers. It is clear from all of this that developers have significant control over the degree with which item transactions will occur within their games.

6. Conclusion

Online games are no longer just games, and item transactions are no longer simple trades between players. As online games continue to evolve, they will act as a marketplace where all kinds of digital economic activities can take place, and item transactions will expand to include more kinds of digital assets.

The gradual shift in perception of digital transactions is important for the Korean online game and digital content industries. Currently, most game developers see online games as just games, with revenue derived from charging access to the game or for premium content. They do not seem to recognize the potential of fields such as item transactions, digital commodities, or cyber banking. Consequently, game developers run the risk of being relegated to a small role in the developing cyber economy, just making online games and nothing more. If developers do not seize the opportunities available in the digital economy, other corporations will.

This perception is also important when it comes to managing the economy within an online game. In-game events commonly give out powerful items to players for little effort, which causes inflation and reduces item value. Inflation can destroy a game's economy, which can spell disaster for the game in the long run.

Online games have the potential to create a complete cyber economy, mimicking, and yet divorced from, the real-world economy. When this potential will be fully realized cannot be predicted, but things are heading down that path even in Korea now. If the cyber economy is ever fully actualized, the industry paradigm will undergo another drastic change.

Chapter 4

The Formation Process of the Korean Online Game Industry

The history of the Korean online gaming industry began in the late 1980s. Though online games would not be widely regarded as a distinct industry by the government and other entities until the late 1990s, the foundations were laid in the previous decade.

Text MUDs (Multi-User Dungeon) were the forefathers of modern online gaming. MUDs, which are devoid of graphics and are played solely via reading and inputting text, first appeared in Korea with the launch of Jurassic Park in 1994. This was just over a decade after MUDs had originally been developed and released in the Western world. Jurassic Park was originally played via PC communication networks such as Hitel and Chollian. The success of Jurassic Park quickly led to other MUDs being marketed, such as Land of Dangun and Ghost Fortress.

Text-based MUDs were a technologically primitive form of online game that nevertheless attracted large numbers of Korean users. Having to take in all of the information from the game via text and input commands in English was inconvenient for users, though. In 1996 Nexon released Kingdom of the Wind, the first MUD to include graphic content (referred to as a MUG, or Multi-User Graphic). In 1998 NCsoft launched Lineage, a full-fledged graphical MMORPG, and the Korean online gaming industry took off on its path of explosive growth.

1. The precursor to the modern online game: Text MUDs

MUD is an acronym which can stand for Multi-User Dungeon, Multi-User Dimension, or Multi-User Dialogue. The differing terms refer to different activities that take place within these worlds. Multi-User Dimension highlights that multiple players inhabit the same world, Multi-User Dungeon refers to a type of game in which the players negotiate a maze or dungeon and fight monsters, and Multi-User Dialogue indicates that the users are able to communicate with each other while online.

MUDs were a revolution in gaming. Offline games, whether on a PC or a gaming console (such as a Playstation), force players to play the game according to the construction of the developer. These games tend to be very linear, with the player progressing through the story until the end. In MUDs, however, the plot is determined by player interaction, cooperation, and competition. The world is alive and changing, and this creates a dynamic environment in which the players live.

MUDs are played via text commands. Figure 4.1 is an example from Jurassic Park.

In a MUD, the player reads the output from the game to determine what is going on around him, and then inputs text commands to order his character to perform actions. MUDs rely on the human imagination, using words to conjure images that immerse players in a fascinating online world. Players work to strengthen their characters in a MUD via a variety of actions, such as killing monsters, collecting treasure, and exploring new areas of the world. In most MUDs, though, the most points are awarded for killing fellow players. Competing with other human players creates a more exciting environment than in an offline game, bringing players back for more and more. Similar to current concerns regarding Internet addiction or chat addiction, a phenomenon of MUD addiction was a cause for worry in Korea in the 1990s.

While Jurassic Park was the first Korean online game, the honor of being the earliest multi-user online game in the world belongs to a

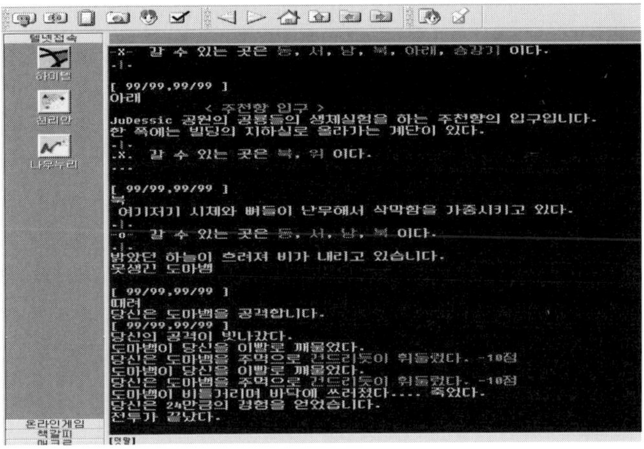

-x- Movement options: East, west, south, north, down, or the elevator.

·.|·

[99/99, 99/99]

Down

<Southwest Sky Entrance>

This is the entrance to the southwest sky where JuDessic Park's dinosaurs are subjected to biological experiments.

There is a staircase leading to the basement at one side.

·|·

·x· Movement options: north, Up.

. . .

[99/99, 99/99]

North

It is strewn with bones and corpses heightening the dreariness.

·|·

–o– Movement options: East, west, south, north.

·|·

The clear sky has given way to rain.

Ugly lizard

[99/99, 99/99]

Hit

You attack the lizard.

[99/99, 99/99]
Your attack missed the target.
The lizard bit you with its teeth. Slight scratch. [98/99]
You brandish your fist at the lizard. -10 points
The lizard bit you with its teeth. It hurts a bit. [93/99]
You brandish your fist at the lizard. -10 points
The lizard staggers and topples. . . . it's dead.
You have earned 24 experience points.
The battle is over.

Figure 4.1 A Scene from Jurassic Park 1

Source: The Jurassic Park Normal-telnet 211.239.157.152

game simply called MUD, installed on the Essex Network in 1980. Essex was absorbed into ARPANet, the international computer research network, which formed the foundations of the modern Internet. MUD's code was originally copywritten, but the source code was shared between students and a culture of sharing the copyrights for MUD code emerged. Developers added their own ideas to the MUD codebase, taking it in new directions and improving on the content. Over time, a large number of different MUDs emerged.

Originally, MUDs were free to players and were created as a hobby or a learning experience by their developers. The number of Internet users was growing, though, and it was only a matter of time until a company tried to commercialize the popularity of online gaming. In 1982 Compuserve, the first ISP (Internet Service Provider), signed a contract with John Taylor and Kelton Flinn of Kesmai Corporation to develop an online roleplaying game in the ASCII-text format. Compuserve soon launched Island of Kesmai and followed shortly thereafter with Megawars 1, a multiplayer game based off of Star Trek. These two titles were the first commercial MUDs.

Most commercial MUDs did not succeed, however, and thus there was never a MUD industry in the West. MUDs failed to appeal to a wide base of users, drawing most of their players from very avid gamers and game developers, and never expanding beyond a niche market. This is what differentiates Korean MUDs from Western ones. Korean MUDs went beyond a simple amusement and became a full-fledged industry.

Video game systems with elaborate graphics, such as Nintendo's Super Famicom (Super Nintendo in the U.S.) and Sony's PlayStation, were dominant in Japan at the time. Famicon, launched in 1980, was followed by Super Famicom in 1990, with the Sega Saturn and Sony PlayStation launching in 1995. These systems provided gamers with a diverse array of titles, such as the Final Fantasy series, the first title of which launched in 1987 for the Famicom and sold 510,000 units. These types of games, with exceptional graphics for the time as well as solid gameplay, felt superior to clunky text-based MUDs to most gamers. This is why MUDs have remained a niche market in Japan, the domain of the avid gamer and never an industry of their own.

The Korean online game industry began to develop at KAIST (Korea Advanced Institute of Science and Technology) around 1990.[1] At the time, Internet access in the country was limited to a few research facilities such as KAIST and the Seoul National University. In the late 1980s some KAIST students were involved with a project to overhaul and revamp DIKU MUD, a game originally developed by students at the University of Copenhagen, with public source code. Only a few students were involved with this project, though. Online gaming remained relatively unknown in Korea at this point.

This changed in 1994 with the release of Samjung Data Systems' Jurassic Park, a MUD based on the Spielberg film. Offered via Chollian and Hitel, two PC communication networks widely used by college students, Jurassic Park was a resounding success and experienced explosive growth. The stated objective of the game was to try and restore the ruined park and contain the dinosaurs, but players reveled in the dynamic world and communication they found in the MUD.

Jurassic Park also served as the introduction of the Game Master (GM), a figure familiar to players of online games today. Present GMs are usually endowed with a broad scope of duties and responsibilities such as customer support, running of in-game events, and monitoring user behavior to report back to the developers. The Jurassic Park GMs, however, were fairly passive and their role consisted primarily of observing players secretly to prevent illicit actions. While monitoring the game world, they would often see players using the world in ways the developers had never thought possible. This unpredictability made the world exciting, and players could always be counted on to find new and unexpected ways to play the game. One type of bug at the time, the so-called "Elevator Girl Obscenity", highlights this:

> "As the game is text-based we created an elevator girl because people need instructions when they enter the elevator. After we installed her we got a report. You could make reports to the developer at the time

[1] Korea Advanced Institute of Science and Technology is a renowned research and education facility in the sciences.

> as well. It was an obscenity report. It turned out that the elevator girl was being subjected to obscenity. I set specific reactions to particular words in a question being posed to her, but the word "obscenity" was not included hence users circumvented the coding I had in mind and engaged in obscenity via hacking. Others stood around clapping. It was something we hadn't even imagined (Oh Choong-yong, the CEO of Samjung Data Systems)."

By 1996 there were over 100 MUDs available to Korean players, with an estimated 200,000 people playing them regularly.[2] Jurassic Park was the most successful, quickly generating over 20,000 visits per day and earning $20,000 in the month of July, 1994. Its growth would eventually reach over $200,000 per month in revenue for Samjung Data Systems.

While they would end up being extremely popular, text-based MUDs had a difficult time making it to market. Korean ISPs (Internet Service Providers) were hesitant to market MUDs in their networks, as they seemed a strange form of entertainment. Executives at the major ISPs of the time had a hard time making sense of them and seeing how they would generate revenue. Oh Choong-yong, the president of Samjung Data Systems, recalls the difficulties faced when dealing with ISPs in the early days of MUD development.

> "It is true today as well but who wants to pay for content? A lot of people refused to pay after viewing content back then as well. I suggested many commercialization ideas which was cause for maintaining an amicable relationship with the people at Chollian, Hitel and Nownuri. Still, when I suggested MUD games, they were apparently taken aback because all they saw were lines of text containing "wham, wham." They later told me that they thought I was insane at the time. Under these conditions I suggested that they trust me blindly and launch the service."

---

[2] This estimate is based on the following. Samjung Data System's Jurassic Park 1 and Mari Telecom's Land of Dangun had a 50% market share in the text-game market. If the user volume is estimated at 50,000 each and the rest of the players playing others games is estimated at 100,000 the number of users of text-based MUDs can be estimated to be 200,000.

Table 4.1 Popular text MUD games

Game name	Game developer	Game plot
Land of Dangun	Mari Telecom	Cooperating with multiple individuals to advance to the highest level of Dangun
Jurassic Park	Samjung Data Systems	Restoring the lost Jurassic Park based on the eponymous Spielberg film
Dragon Land	Digital Impact	A fantasy world wherein the Oopsala Empire is defended from monsters
Aurora Camp	Dyne Media	A North Pole excavation is depicted in the Social Mud form
Time Traveler	Hanme Network	Exploring the world in the passage of time
SF 1999	Hyunmu Soft	A game based on the space theory in outer space

It was difficult to convince the Korean ISPs to offer MUDs as part of their network, but shortly after launching, MUDs became the primary source of revenue for most service providers. The number of players who could simultaneously log in to Jurassic Park was initially capped at 300, but this cap was quickly reached and exceeded, and the game would eventually capture a user base of over 100,000. By charging an access fee of roughly $0.01/minute with a 6:4 or 5:5 profit split with the ISP, Samjung Data Systems' monthly revenue surpassed $100,000. The success of Jurassic Park led every Korean ISP to offer MUDs to their users within short order. Table 4.1 is a summary of the most popular commercial MUDs at the time.

Table 4.2 highlights the three major ISPs of the time and their top service offerings. Notice the prevalence of MUDs in the rankings.

As Table 4.2 shows, MUDs were the primary source of revenue for each of the three largest Korean ISPs. With this explosive growth, however, came a budding conflict centered around revenue distribution between MUD developers and the service providers who hosted the games.

Table 4.2 Top three PC communication service providers service offering rank

	KT Corp (Hitel)		Dacom (Chollian)		Nowcom (Nownuri)	
Rank	DB	Type	DB	Type	DB	Type
1	Jurassic Park	MUD	Jurassic Park	MUD	Land of Dangun	MUD
2	Oksori	Song info	Koscom	Stock market info	Jurassic Park	MUD
3	Koscom	Stock market info	Time Traveler	MUD	Time Stranger	MUD
4	Our Star	Entertainment news	Next-generation archive	Archive	Screen	Movie info
5	Tohjungbigyul (Fortune)	Fortunetelling site	Internet	Internet access	Window Story	Computer instruction
6	Recruit job info	Job site	Saju Doctor	Fortunetelling site	Sports Seoul	Provides newspaper articles
7	Cinetel	Movie info	Screen	Movie info	Billiards	Sports game
8	Sports Seoul	Provides newspaper articles	Sports Seoul	Provides newspaper articles	Internet	Internet access
9	Jaehyun System	Corporate Advertisement	Quiz Heaven	Quiz game	Quiz Heaven	Quiz game
10	Job Data Line	Job site	Employment info	Job site	Computer Paradise	Computer instruction

Source: PC World (April 1995 issue) excerpt

When MUDs were first offered online, the revenue split between the developers and the ISPs was typically 9:1 or 8:2 in favor of the game developers. As MUD revenue grew and the service providers saw that these games represented their largest source of income, they saw an opportunity to bolster their bottom line and the profit split grew more contentious.

ISPs negotiated with MUD developers from a position of strength, as there were only a few options available for MUDs to be hosted where they could reach large numbers of users. Without an ISP to provide hosting and payment collection services, a MUD had no market. MUD developers had no choice but to abide by harsher terms imposed by ISPs. Soon it became common for an ISP to keep 40%, 50%, and in some instances even 90% of a MUD's profits.

Despite the explosive commercial success of MUDs many game developers, including industry leaders Samjung Data Systems and Mari Telecom, faced financial difficulties due to the new arrangements with their service providers. Oh Choong-yong, CEO of Samjung Data Systems, recounts this period of his company's history.

> "We selected PC communication service providers as a vehicle for fee collection. While we had no means of directly collecting fees, PC communication service providers charged a monthly usage fee. At first they suggested an 8-to-2 split; 8 for us, 2 for them. At the time the government was implementing a CP-promotion policy which required profit sharing with content-providers. Simply put, if 100% of the profit from CP service were allotted to the CP provider, 95% of them made less than $1,000. That's where things stood back then, and market conditions were such that if they agreed to a 5:5 split they would all go under. We began with 2:8 and 1:9, like everyone else, but when they popped the lid, their attitude changed dramatically and suggested 6:4. They claimed they never imagined it would take off like that. Once, we made 10% of whole revinue of PC communication service provider 10%. And we began on footing advantageous to us, but the terms grew increasingly severe. But our hands were tied. The ball was in their court."

2. MUG (Multi User Graphic) game development and the spread of users

The Korean online gaming market was advancing in two distinct directions from the commercialization of MUDs in 1994. The first was text-based MUDs, such as Jurassic Park and Land of Dangun. The second was modem games played directly between users, such as billiards and car racing. Both types of games were offered by Internet service providers of the time, but their attributes and gameplay were very different. MUDs were hosted entirely on the ISP's servers and relied on text for their gameplay. Modem games, on the other hand, were able to include graphics as the games themselves actually ran on the user's PC, and the ISP only served as an intermediary to connect the game's players. Another difference is that while dozens to hundreds of players could simultaneously play the same MUD, only a few players could play in the same instance of a modem game.

Because they were persistent worlds and could host large numbers of players at once, MUDs were based far more on community than other games. The community that would spring up around a MUD was generally considered the defining factor of that particular game, and this trait held true through the development of MUG games like Kingdom of the Wind and Lineage.

Modem games can be seen as the halfway point between traditional offline PC games and full-fledged online games. A large persistent user base such as MUDs had did not develop around modem games, but they were successful at attracting new gamers to the genre. While MUDs were the domain of the avid gamer, many casual gamers played modem games. These games would eventually evolve into Web-based board games such as Han Game.

In December 1995, Nexon launched the beta version of Kingdom of the Wind via the ISP Chollian. The full commercial version would follow in April 1996. Nexon was the first domestic game developer to implement graphics in a MUD. Taewool, a company founded by KAIST graduate Cho Hyun-tae, developed a MUG named Foundation and launched it in June 1995 with the ISP Unitel as its host.

Nexon and Taewool were pioneers for their time, but they faced many difficulties in bringing their games to market. Gaming was not yet viewed as a full-fledged industry, and thus gaming companies struggled to recruit able developers. Many developers also failed to grasp the concepts and principles behind the MUG, as the games were split between a server to run the game and connect users, and a client on the user's PC to connect to the server with and display the game's graphics. Jake Song recalls the industry concensus at the time that a game with a client size of 1.4 megabites (the size of a single floppy disk) was a crapshoot at best.

> "At the time most industry people remained skeptical saying "the speed is insufficient for transmitting graphics in MUD." But why transmit all the graphics? Pictures are already downloaded, and with 1 signaling for one cell up and 2 for one cell down, transmitting one byte is sufficient, perhaps resulting in a lower volume being transmitted. I thought bandwidth wouldn't be a problem, and I was vaguely convinced that graphic MUDs will work wherever text MUDs work. And when development begins no developer tailors to the current specs. Rather, they tailor to the maximum current specs because two years later, that will be the norm. Or they could shoot for specs which don't even exist yet, harboring the vague hope that it should surface after two years. That was the case for us as well."

Most game developers shied away from following Nexon and Taewool into the realm of graphical online games. Industry forecasts for the success of MUGs was dismal, and most game developers did not have programmers on staff who had the skills to create this type of software. Jake Song has this to say on that phenomenon.

> "Firms which developed text MUDs basically lacked the technology. Frankly text MUDs hinge on words. No one possessed in-house technology aside from Mari Telecom because they were merely turning out duplicates. Some brought over DIKU MUD and some LP MUD, but they were all duplicates. There was no technology to

> make graphic MUDs, and it was pure nonsense to convert to graphic-based games at a time when the industry consensus expected graphic MUDs to fall through."

Kingdom of the Wind, the trailblazer of graphical online games, was not a commercial success in its early stages. When it was first released, the number of concurrent users stagnated in the low hundreds. Without a PvP (Player vs Player) or castle siege system, Kingdom of the Wind had no content to promote inter-guild strife or intra-guild cooperation. These elements were what drew players to MUDs and what kept them playing over the long term, and without them Kingdom of the Wind had trouble attracting and keeping players. Until 1998 Nexon was losing money on Kingdom of Wind, and turned to performing IT and web development services to stay afloat and fund its game.

Despite its unprofitability, Kingdom of the Wind was important to the game industry as a whole, serving as a prototype of the first graphic-based game in a predominantly text-based industry. In the mid 1990's, the viability of a graphical online game was not yet established, and Kingdom of the Wind provided proof of both the technological capacity to build such a game and the existence of prospective users. Unlike text-based MUDs, which were populated by hardcore gamers, Kingdom of the Wind succeeded in drawing in ordinary users with its graphical appeal.

Kingdom of the Wind was also the introduction of the real-time graphical roleplaying game (RPG) game system, in which players create and build up a character in a graphical online world. It was a pioneer in the Korean game market where the RPG was still a novel concept. Unfortunately, it was not sufficiently developed to touch off the Korean online game industry explosion, which would have to wait for the arrival of Lineage.

3. The emergence of Lineage and industry expansion

Lineage is the game which would become the foundation of the Korean online game industry and which even today, 10 years after its

Table 4.3 Rising trend in Lineage service user number

1997.3	NCsoft established
1998.9	Lineage domestic commercialization
1998.12	First domestic online game to exceed 1,000 players logged on simultaneously
1999.12	First in Korea to exceed 10,000 players logged on simultaneously
2000.12	Number of players logged on simultaneously exceeds 100,000
2001.12	First in the world to exceed 300,000 players logged on simultaneously

Source: <http://www.ncsoft.net/>

launch, still boasts over 130,000 concurrent users. As shown in Table 4.3, Lineage broke the 1,000 concurrent user mark 4 months after launch, 10,000 within one year, and would eventually set a world record 2 years after launch with 300,000 concurrent users. Released at a time when Blizzard's Starcraft was dominating the Korean game market, NCsoft's Lineage triggered a massive influx of players to online games and solidified the online game business model.

Lineage used the basic RPG system of classes and character levels, but emphasized fast action to hold players' interest. Users could choose from 5 character classes: monarch, knight, elf, mage, and dark elf. All classes distinguish between male and female characters, thus players actually have 10 different character types to choose from. Boasting complete inter-guild warfare and castle siege systems, Lineage presented the most sophisticated online world yet seen in a game.

One of the major factors contributing to Lineage's success was that it was targeted towards ordinary users, not hardcore gamers. The game could be controlled solely with a mouse, and the interface was simple and easy to understand. Lineage was a very approachable game for new players, which aided in its rapid expansion.

The defining characteristic of Lineage was that it acted like an actual online society, not just a game. Instead of the developer hosting events and guiding the story of the game, the players are the ones who decide in which direction the game will go. In Lineage there exists a type of guild known as a "blood alliance", in which players swear fealty to a monarch and fight for a common cause. Blood

Figure 4.2 Scene of a siege in Lineage

Source: <http://www.lineage.co.kr/linweb/lin_main.asp>

alliances draw players together into tight-knit communities, and these alliances have contributed significantly to extending the shelf life of Lineage in the market.

The rapid spread of Internet cafes in Korea was another instrumental factor in the success of Lineage. Internet cafés served not only as a site to access the game, but also provided marketing to help grow the game's base of users.

The expansion of Internet cafés offered NCsoft an excellent way to collect revenue. Until 2001, most of Lineage's revenue came via Internet cafés, as shown by Table 4.4 which highlights NCsoft's sales by sector. In 2000 Internet café sales accounted for over 70% of NCsoft's annual revenue, more than 3 times the revenue that came from individual users on home computers. As time has gone by this balance has shifted, and as of 2003 revenue from individual users has eclipsed revenue from Internet cafés. This is a result of the PC industry producing faster and cheaper computers as well as improved residential access to high-speed Internet connections, both of which

Table 4.4 Sales by sector (Unit: 1 million dollar)

Sector	2000	2001	2002	2003	2004
Internet Cafés	39.386	61.370	53.331	48.005	74.600
Individual	10.738	43.558	74.726	88.603	133.788
PC Communication	4.479	8.336	4.680	1.681	0.547
Royalty	1.296	9.327	21.873	28.248	37.959
Total	55.900	122.593	154.610	166.537	246.894

Source: <www.ncsoft.net>

contributed to users connecting to Lineage from their home computers rather than going to an Internet café to play.

Nexon's Kingdom of the Wind also generated revenue via Internet cafes, but such sales represented only a small part of Nexon's revenue. The Internet café revenue model might have been originated by Nexon, but Lineage was the game that put the model on firm footing and contributed to the growth of the Internet café industry.

Lineage's success brought about many competitors. Game developers saw the huge profits NCsoft was earning and rushed to create their own graphical online games. Developer after developer announced that they were building a new MMORPG, and the number of MMORPGs released surged from 9 in 1999 to over 100 in 2000. It seems safe to conclude that most of these games were inspired by the success of Lineage.

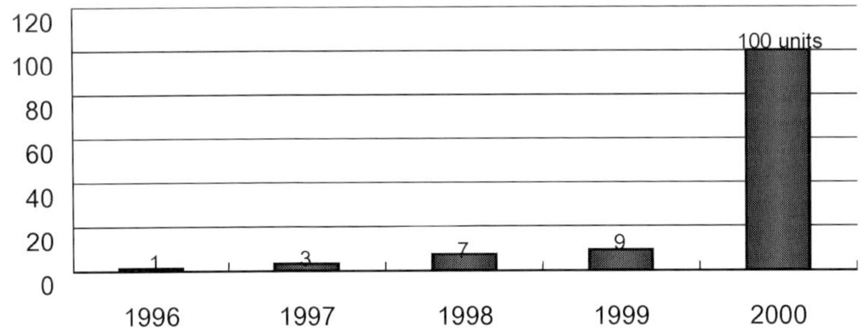

Figure 4.3 Online game release trend

Source: <www.gamemeca.co.kr>

Lineage also create a new revenue model for online games known as "item cash transactions", in which players purchase in-game items with real-life currency. This was a switch from the older model of paying a per-minte, per-hour, or per-month fee to connect to a game. Lineage currently boasts the largest number of item cash transactions of any MMORPG. According to ItemBay, an item cash transaction intermediary website, 40% of all transactions involve Lineage.

Primarily from the strength of Lineage, NCsoft's sales topped $100 million in 2001, and as of September 2005 sales have already surpassed $800 million for the year. After seeing the incredible success of Lineage and the meteoric rise of NCsoft, game developers have concentrated on creating their own Lineage-type MMORPGs. Lineage has become the benchmark for Korean game development studios which lack extensive online game experience of their own and have borrowed elements such as game systems, graphics, world view, user service, and customer support from Lineage.

4. The emergence of game portals

With the introduction of "Han Game" in the late 1990's, another shift took place in the online gaming market, which up to this point had been dominated by MMORPGs. Han Game offered more traditional games such as Go-Stop, poker, and chess over the Internet. While the user base of MMORPGs like Lineage was generally limited to a particular demographic group (young males), Han Game's titles were accessible to anyone. Women and older adults, who tended to avoid games like Lineage, realised there was a game they could enjoy on the Internet.

Han Game was founded in September 1998 and launched its online service in December 1999. Its site was an immediate success, with a registered membership of over 1 million users within 3 months and daily page views over 10 million. Han Game quickly became one of the leading online game sites.

Han Game was founded by former employees of Unitel, Samsung's Internet service provider. Well aware of the limitations of direct PC communication as well as the awesome potential of the

Internet, Han Game decided to build its site with an eye towards the future. They predicted that in the near future the prevalence of the Internet would render connection to a communication network via modems unnecessary and that the direct-connection PC communication industry would become obsolete. The Internet's prospective user base also consisted of every online user, rather than being limited to the registered users of a specific service provider. This gave the Internet an almost infinite potential for growth.

In its early days, Han Game adhered to the typical Internet business model of "Internet as medium". Their model was to collect users via websites and generate revenue by selling advertising on those websites. One difference between Han Game and the plethora of other Internet startups attempting to do the same thing is that Han Game formed a registered user base with its games. That distinction is precisely what made today's Han Game possible.

There was careful reasoning behind selecting games as the method with which to secure registered users for their sites. Han Game believed that games had a high possibility of successfully turning commercial, as opposed to other online content (such as news) which would likely always remain free. Selling advertising was the central source of revenue in their business model, but they believed that if they needed to, they could commercialize their games and earn a profit in that way. Namkung Hoon, a Han Game executive, offers the following comments on this.

> "We conducted research on our own as U.S. led the Internet market... and discovered that there were only two ways of making money on the Internet at the time: porn and game sites. Games were the first successful domestic commercial sites as well. We thought that we should at least be able to get by with going commercial, that it might be possible to make users pay to run the company. Frankly we didn't have high expectations riding on going commercial. At the time we still believed that instead of user fees accounting for a majority of our sales, it would turn into a medium channeling advertising from TV networks, newspapers and radio. Where as TV or newspapers are unilateral communication, the Internet is bilateral hence the

> advertising potential of the Internet would be far larger than those of other media. We approached it with a premise that going commercial won't at least topple the company."

When the dotcom bubble burst in 2000, Internet companies faced an almost total loss of advertising revenue. Most of them would collapse as they had no other source of income. Han Game faced the same loss of advertising revenue and realized it could not survive without going commercial. The executives at Han Game were concerned with the ramifications of going commercial, though, such as user protests and losing users altogether.[3] Their answer to this was partial commercialization.

In November 2000 a chat service called "Say Club", serviced by Neowiz, launched partial commercialization a few months before Han Game. Via the cellphone microtransaction payment method, Say Club sold Avatar items.[4] Say Club was successful in doing so, and Han Game introduced a similar form of commercialization.

Han Game began their commercialization in March 2001 with a twofold approach. The first method was via item sales, selling items for games such as Tetris, Go-Stop, and Yut at prices ranging from $0.10 to $0.60 per item. The second method was by charging a monthly usage fee of roughly $8.00 which allowed for unlimited play of all Han Game's titles. The strategy behind this was to collect fees via item sales from heavy users who spent large amounts of time on the service while allowing light users to play for free.

> "We figured someone who plays more than a few rounds of games is a heavy user, and he has no choice but to play, and we concluded

---

[3] According to a survey conducted on 20,000 Internet users by Shimani in late 2000, 54.6% of the respondents replied 'Internet should be free' (Chosun.com "Nothing's Free Declaration/Paying for Internet Content," February 2001). According to a face-to-face survey conducted by Han Game in January 2001, 62.5% of the respondents replied that if a site goes commercial, their decision to use the site will be determined by price and content quality.

[4] Avatar refers to the incarnation of a Hindu deity, and is a character which symbolizes users in the game. Users first select one's character's physical attributes on an Avatar business site, and incorporate wigs, wardrobe and shoes to make it into an entity resembling the user. Wigs, wardrobe and shoes which are called "items" are all paid items and contributed to the spread of the cellphone pay method.

that the light user will stop playing if we charge them to play. That's why we allowed light users to play for free and with heavy users who play our game extensively and take up server space and create a lot of traffic... we thought 'the amusement they get out of it has to be worth at least $0.3 a day to them.' That was our position."

Despite some user protests, the commercialization of Han Game was a success. On the first day of going commercial they recorded sales of $50,000, and monthly sales were quickly over $700,000. Going commercial did, however, have side effects, including losing the teenage player base. Teenagers tend to be very sensitive to the price of online content, and with Han Game going commercial most of them switched to Net Marble, a free game site.[5]

The successful commercialization of Han Game is telling not only for the online game market, but for the Internet as a whole. Very few advertising-driven sites were successful in their efforts to commercialize their user base following the dotcom implosion. At a time when users were opposed to the idea of paying for Internet content, Han Game's success demonstrated a model that could be applied to other businesses, and indeed many other Internet companies would follow in Han Game's footsteps in their commercialization plans.

Han Game's most significant achievement as applies to the online game industry as a whole was not its commercialization, however, but its widespread user appeal. Online games had previously been the domain of MMORPG players, and Han Game changed that and opened up online gaming to a much broader base of casual players. Their games were widely known, such as Go-Stop and billiards, and were thus very accessible to and popular with female and older adult users. Han Game's success introduced a large number of Internet users to the concept of playing games online.

Han Game also introduced a new model for online gaming, that of Web-based board games. They took already existing and relatively simple games like checkers and chess, put them online, and managed to create a large user base of players for these games. In Han Game,

[5] Net Marble is one of Han Game's competitors primarily offering Web board games.

anybody can log on and access a game quickly and easily, as well as meet other players and engage in conversation. In addition, premium services such as avatars are provided. Han Game successfully developed a large Internet community around its titles, going above and beyond mere games.

5. The emergence of casual games

In 1999 another transition took place in the Korean online game industry. Fortress 2, a casual game based around tank warfare, was released by CCR. Fortress 2 pioneered casual games as a new genre in the online game market. Its simplicity and ease of use attracted light users, such as women, and it could be quickly picked up due to its simple user interface. Fortress 2 was remarkable for greatly expanding the online gaming market and introducing new players to the genre.

CCR, the developer of Fortress 2, was founded in 1995 by students of Hanyang University's computer science program. Yoon Seok-ho launched the company with 5 friends while he was a member of the Samsung Electronics Software Membership, an association of software developers. CCR was not originally formed to be a game developer, however. Its early business was systems integration (SI) and other general IT work. In 1995 CCR constructed the POSCO Intelligent Building System, and in 1996 they handled primarily Internet-related SI/IT work such as developing Hansol Telecom's free Internet advertising system.

CCR's development ability began to attract attention in 1996 with the release of the web browser "X2 Web". X2 Web was a browser which changed its appearance to display corporate logos, trademarks, or product advertisements according to the URL it was currently accessing. While not related to games, this browser brought CCR into the limelight and made its name familiar in industry circles.

CCR began to develop games in 1998 and became a full-fledged game developer with the launch of Fortress 2 in 1999. Having developed SK Telecom's PC communication service "Nets Go" while working as an IT firm, CCR first offered Fortress 1 via Nets Go.

At the time, Nets Go was experimenting with many different types of content and Fortress 1 was more popular than the daily news. Having confirmed the opportunity in online games via Nets Go, CCR independently developed Fortress 2 and launched open beta in late 1999, which attracted an incredible 3,500,000 registered users within 6 months.

Fortress 2 is a game of tank warfare in which each player controls a tank and attempts to destroy the opposing tanks or force them off the edge of the map. Fortress 2 can be controlled with just the space bar and arrow keys, a very convenient control system which made the game widely accessible to any Internet user and drew a player base of diverse age groups. Fortress 2's user base varied from age 10 to 80 with 40% of the players being female, something very unusual for an online game at the time.

Fortress 2's strength was that while it was simple to learn, the game was deceptively complex. While it seems like a simple point-and-shoot game, the location and angle of the opponent's tank played a major factor in the effectiveness of a shot, and the game required diverse strategies to be successful. Matches could consist of one-on-one duels to four-on-four battles. With every shot fired, tips and commentary would inundate the chat window on the screen. Yoon Seok-ho, president of CCR, describes Fortress 2 as follows.

> "Koreans prefer games wherein angle and strength can be controlled. Billiards is a prime example. It has simple rules but no one refers to it as a simple game. This is what we wanted. We began developing with a complex game with simple rules in mind."

Twelve different models of tank are featured in Fortress 2. Each tank is distinct, with strengths and weaknesses, and this adds to the game's strategy. While no tank is the best, each tank must utilize different strategies depending on the tank the opponent has chosen.

Casual games such as Fortress 2 face a challenge when trying to go commercial. MMORPGs such as Lineage have a well-established monthly fee system as well as item transactions, and most of their users put hours upon hours into the game and thus felt it worthwhile to pay a monthly fee. Most of Fortress 2's players were light users, however, and thus a monthly fee was not a viable option.

After examining different pricing options, CCR opted to charge a fee to Internet cafes in order to access Fortress 2 rather than to individual players. The game was updated to Fortress 2 Blue and beginning on January 2, 2001 CCR charged a monthly usage fee of roughly $5 per computer for Internet cafés with less than 40 computers (see Table 4.5).

CCR stated that this fee was to reduce the loss they were taking to operate their servers and to keep their promise to their registered users of not going commercial. They also promised to purchase an additional 35 servers exclusively for Internet café connections to provide enhanced service. The primary sites for accessing online games at this time was via Internet cafes, and charging the cafés themselves rather than the users caused a conflict with café owners. Internet café associations visited CCR and held rallies

Figure 4.4 Tanks of "Fortress 2"

Source: <http://fortress2.x2game.com/>

Table 4.5 Pricing plan for internet cafés as of January 2001 — according to number of IP units

Less than 5 units	6–40 units	41–60 units	61–80 units	81–100 units	More than 101 units
$40	$217	$326	$392	$435	$500

Source: CCR internal documents

to protest the payment model, claiming that their cafes had been saddled with fees that should rightfully have come from the individual users.

This conflict with Internet café owners was eventually defused with the intervention of the Ministry of Culture and Tourism. The Ministry served as an intermediary for negotiations between CCR and major Internet café associations, which eventually led to a cooperation agreement being signed in March 2001. The agreement stipulated a maximum fee for access to Fortress 2 Blue that Internet café owners felt more comfortable with.

Following the agreement, CCR's fees to Internet cafés for access to Fortress 2 Blue dropped significantly, and CCR has offered cash item transactions as an alternative to help generate revenue. Items available in Fortress 2 Blue include unique player names using special text, color chatting, establishing secret chambers, unlimited connectivity, and others.

Games such as Kingdom of the Wind, Lineage, Han Game's board games, and Fortress 2 highlight a massive expansion in the graphic online game market in Korea. Online gaming is no longer limited to RPGs, but includes diverse genres from board games to sports games.

Online games now attract massive budgets as game developers release the profit potential they hold. Huge titles such as NCsoft's "Lineage 2" and CCR's "RF Online" are dominant in the Korean online game market, and new types of games such as the sports game "Pangya" and board game "Pimang" are gaining popularity as well.

Table 4.6 Internet café pricing plan as of November 2004 — according to number of IP units

1 unit	2 units	3–40 units	41–60 units	61–80 units	81–100 units	More than 101 units
$40	$80	$99	$150	$180	$200	$230

Source: CCR internal data

6. Formation of the innovative human resources: Online game development workforce

Online games require more diverse technology than offline games, and the lack of developers possessing the requisite skillsets was a challenge facing the industry in its nascent years. Online games require server and network technology to be combined with game technology to create both client and server programs. Server program run on a game's central server, while client programs are installed on the end user's PC. The server program needs to accommodate thousands to hundreds of thousands of simultaneous users, and the programming skill required to create such a program is extremely advanced.

An important attribute of the Korean online game industry is the entry of skilled developers into the field. Engineers who majored in computer science or networking in college joined the industry and became developers, bringing their server networking knowledge to their companies.

This stands in contrast to Japan, where engineers rarely sought jobs at game developers. At major developers such as Nintendo and Sega in-house training exists, but until online games became a major industry, no game developers trained their employees in network technology. This left Japanese developers at a disadvantage when the online gaming boom began.

To see how such a skilled workforce entered the game industry in Korea, we can examine the career of Jake Song, the developer of Lineage and Kingdom of Wind.

Kim Ji-ho, a KAIST graduate and long-time MUD developer, was working on Kit MUD, an improved version of the University of Copenhagen's DIKU MUD. At the time, Jake Song was a Master's student at KAIST and caught a glimpse of Kit MUD while it was in development. Song became enamored with Kit MUD and began working on it, learning different types of MUD source code as well as about the networking and server technology required to access a MUD online.

After completing his Master's Song enrolled in a PhD program at KAIST, but withdrew after a year and a half to take a job at Hansoft.[6] Song aspired to develop games, but was repeatedly turned down by Hansoft, which had no interest in creating video games. Song continued to push online gaming to Hansoft and was finally allotted one month to work on online game development. He partnered with Lee Hee-hang (Vice President at NCsoft) and with Jake Song on the server end and Lee on the client end, development of the game engine looked promising.[7]

They named their engine DOOMVAS (Distributed Object Oriented Multimedia Virtual Active System), and it was designed solely to create online games. Song explains the reasoning behind the transition to graphical game development instead of an extension of existing text-based MUDs as follows.

> "It was around 1991. I was playing a Kit MUD game, but I got lost frequently. And I thought it might be easier for users to understand if the game was executed with graphics instead of text or if something like maps indicated locations or conditions around them. However, at the time many voiced skepticism on graphic game development because with the maximum speed of PC communication being too slow at 9600 bps, playing MUGs while transmitting mass information was unthinkable."

While developing DOOMVAS they met with IBM Korea and secured a venture capital investment. IBM Korea was excited by the prospect of the game they had in development in which large numbers of users could play simultaneously via the Internet. At the time IBM Korea had an affiliate program which provided for collaboration with startup companies. IBM would provide the technology, funding, sales network and marketing and the startup firm developed the product and shared the profits.

---

[6] Hansoft is a firm developing a Korean word processing program.

[7] Both are engineers who developed NCsoft's Lineage.

In September of 1994 IBM Korea was soliciting ideas from its employees for investment opportunities. They suggested investing in multimedia content development, and $50,000 was earmarked for DOOMVAS with IBM Korea preparing to collaborate with Hansoft.[8] This deal would never materialize, though, as Song left Hansoft in December 1994.

Song left Hansoft to establish Nexon in cooperation with Kim Jung-ju, a Seoul National University graduate. Nexon's goal was to focus solely on online game development. Nexon's first game, as discussed previously, was Kingdom of Wind. Jake song relates the production process of Kingdom of the Wind below.

> "We didn't have a clear-cut plan from the outset. We thought it would be nice to have an original work the game was based on. As we were both engineers, there were limitations our scriptwriting. While searching for something that could serve as an original script from works of fiction, we thought graphic novels might work well. But we ultimately gave up as it was a crapshoot whether popular comic authors would want to collaborate with us. Instead we asked Kim Jin, a manhwa author to do it. As he was a former illustrator at Hansoft we thought we should seek his help."

IBM Korea was still interested in Song's work, and they signed a joint-development agreement with Nexon when Song left Hansoft. IBM invested over $50,000, provided another $90,000 worth of server equipment needed for online game development, and agreed to act as Nexon's global marketing agent. Development took longer than anticipated, however, and Nexon faced a cashflow crisis in 1995. In December of that year IBM Korea provided another $45,000 in funding and gave Nexon the first Korean website, that of Hyundai Motors, as a project to provide some revenue to get by.

In October of 1995, 2 months before the additional round of investment from IBM Korea, Jake song left Nexon due to differences with co-founder Kim Jung-ju. His terms of departure stipulated that

---

[8] From the interview with Ko Jae-bum conducted on April 11, 2002 from 3:00 p.m. to 5:30 p.m.

"while giving up on 50% of Nexon shares he is free to develop new games employing the Kingdom of the Wind source code, and after 1997 games developed in such a manner can be launched."

After leaving Nexon, Song took a position with an ISP named Inet and began development on Lineage in September 1996. He was given one year to create the full game. Before he could complete his project the Asian financial crisis of 1997 struck, and difficulties caused Inet to scrap many of its projects, including Lineage. Discouraged by yet another failure in trying to create Lineage, Song left Inet and looked elsewhere.

While he was looking for another job, Kim Taek-jin, the CEO of NCsoft, suggested that Song come on board. Jake Song began working at NCsoft in late 1997, continued working on Lineage, and the game finally launched in September 1998. One year and 10 months after launch Lineage recorded 3 million registered users, leading the charge of the Korean online game market's amazing growth story.

Chapter 5

Factors Promoting Growth of the Online Game Industry

The Korean online game industry has grown at a spectacular rate — a rate far faster than that of other major nations such as the U.S. and Japan, both of which have an established console game industry predating that of the Koreans. What accounts for the explosive growth of online gaming in Korea that is above and beyond the rest of the industrialized world?

It is commonly believed, especially by foreign governments and game developers, that the primary explanation for the growth of online gaming in Korea is intensive government funding. The fact is the Korean government did not begin to fund online gaming until the industry was well on its way to success. If not for government funding, what then was the key to success for the online game industry?

There are two factors that contributed significantly to the rise of online gaming in Korea. First is the formation of the workforce of skilled programmers and network engineers required to develop online games, as discussed in Chapter 4. The second is the timely emergence of infrastructure to support the industry, such as mobile small-amount billing and Internet cafés.

Thus in this chapter, we will examine factors which aided in the development of the online game industry. We will examine early market conditions surrounding the offline game industry, infrastructure developments such as Internet cafés, broadband Internet access, and mobile small-amount cell phone billing, and government policy such as an adapted military term of service that helped supply the workforce the game industry needed and policies that fostered competition in the network communication infrastructure.

1. User's path dependency: Absence of the console game market

The history of the offline console game market significantly influenced the spread of online gaming. In markets where gamers had become acclimated to offline gaming, either on video game consoles or PCs, online gaming had a harder time establishing a foothold. The phenomenon of users who have grown acclimated to one game platform and are unwilling to move to another platform is referred to as "Path Dependency". This term explains how previous product selection or education will influence and even restrict current decision-making (David, 1985; 1997).

The QWERTY keyboard is a prime example of path dependency. In 1867, C.L. Sholes manufactured and sold practical typewriters. These typewriters had a drawback, however. When the user typed too fast, the type bars connected to the keys would become tangled. In 1873, Sholes collaborated with Remington to develop a typewriter with a QWERTY keyboard, which resolved the issues facing the first model.

As time went by, electric typewriters were developed, which solved the tangled type bar problem of the original mechanic typewriter. A more efficient DSK (Dvorak Simplified Keyboard) was invented and introduced with the goal of replacing the QWERTY keyboard, but the new keyboard was rejected by consumers who were unwilling to switch and learn a new style of typing. Organizations such as the Typewriter Training Association worked to keep the QWERTY keyboard dominant. This is a prime example of how path dependency can delay or even completely impede the spread of superior products or technology.

Path dependency has affected the spread of online games in a similar manner. Online gaming is expanding most rapidly in parts of the

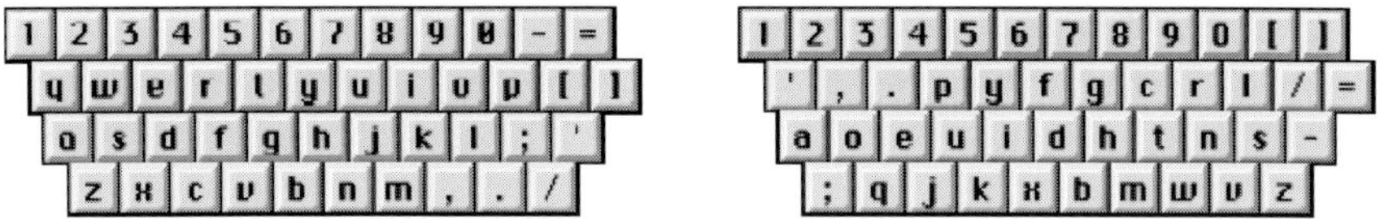

QWERTY keyboard | Dvorak keyboard

Figure 5.1 QWERTY keyboard and DSK keyboard

Source: David (1985)

Asian market where offline console gaming has not yet made strong inroads, such as Korea and China. Markets with a long history of offline console game playing, such as the U.S. and Japan, are seeing much slower growth in online gaming (Wi, 2004; 2006a; 2006b).

While offline games are primarily played using game console controllers, online games are played with computer keyboards. Users who have grown used to playing games with controllers can feel inconvenienced by having to use computer keyboards to play games. Once users have become acclimated to using a few simple buttons and a joystick to control a console game, the prospect of using a keyboard with more than a hundred different keys can seem intimidating and confusing, especially if the player does not already type.

Offline console gaming never took hold in Korea to the same extent it did in the U.S. and Japan. Accordingly, Korean gamers became immersed in online text-based MUDs to a much larger degree than in the other two aforementioned nations.

The Korean console game market began to develop in 1989 when Samsung and Sega joined forces to market their 8-bit "Aladdin Boy" console. At the time, Samsung Electronics had 15 domestic software developers working on game development. Other companies soon entered the console gaming market as well. Hyundai Electronics collaborated with Nintendo to create and market 8 and 16-bit game consoles, LG Electronics invested $10 million in the development of the 3DO console, and Samsung released the 32-bit Samsung Saturn console in October 1995 (Roh and Wi, 2007).

Not for lack of corporate effort, though, the domestic offline game market was stagnant. The Samsung Saturn, for example, sold a mere 30,000 units. Two factors accounted for the poor sales of domestic game consoles.

First is the lack of available titles for the available consoles. Domestic game developers were in their fledgling stages and were unable to turn out enough titles to satisfy consumers. The importing of Japanese games was banned at the time, and due to limited development ability, the creation of Korean versions of Japanese games was not viable.

The second reason why domestic game consoles never took off in Korea was because of competition with contraband Japanese game consoles. The Korean government banned imports of Japanese consoles to protect the domestic market; however, there was an underground market for smuggled Japanese consoles. While the Samsung Saturn sold for $300, a smuggled Japanese equivalent was able to dodge the sales tax and was available at only $200. This underground trade in Japanese consoles prevented Korean developers from expanding the domestic game market.

In 1996, LG Electronics decided to cancel production of the 3DO, and Samsung Electronics followed LG in exiting the console market. With the conglomerates leaving the industry, the Korean console game market became dominated by Japanese and American products, such as the PlayStation and Xbox. Korean software developers attempted to shift production to developing software for Japanese consoles, but were impeded by steep royalties and expensive development engines.[1]

These conditions meant that console gaming in Korea was limited to avid gamers and never spread to the common consumer market. The size of the Korean console game market (including both software and hardware) was just $12.5 million in 2000, diminutive compared to the $116 million in revenue seen by the PC game market in that year.

The weakness in console gaming was a boon for the nascent online gaming industry. In Japan, the strength of the offline gaming

[1] Sony officially launched PlayStation in February 2002 and selected domestic software developers like Phantagram and Softmax began developing products for PlayStation.

market and high consumer loyalty combined to delay the transition from offline to online gaming. In Korea, however, an entrenched console gaming industry to serve as a powerful rival did not exist.

2. Software piracy

Piracy, normally viewed as negative in the console game market, actually has some benefits for online gaming. Console and PC games can be copied and pirated with relative ease, but because of the interactions with developer-run servers, online games are far harder to pirate. When large amounts of pirated software are available, game developers' sales and profitability fall, indicating that they are unable to recoup development costs and their efforts to create new titles are deterred.

In Korea, PC and console game developers both ended up promoting a switch from offline to online gaming. Developers facing the challenges of software piracy were eager to switch to creating online games due to their difficulty to pirate. Game developers had been plagued by piracy for years in Korea, in both PC and console gaming, and they were eager for a more secure business.

In the console industry, consoles themselves are designed to prohibit the use of pirated game software. But hackers developed chips which circumvented this protection and enabled the use of duplicated software. Consoles with these chips were sold via specialty electronics retailers to game enthusiasts, who then purchased duplicated software from Taiwan or Hong Kong instead of more expensive authentic software.

The PC game industry also had its share of piracy problems. In the early 1990's, games were distributed via floppy disk, and piracy was a simple matter of copying disk data. The data medium later changed to CD-ROMs, but even then the piracy problem was still prevalent.

The Korean government recognized piracy as a problem and reformed piracy law in the early 1990's. They tried to crack down on software piracy in the mid 1990's, but the problem remained, exacerbated by the ease with which pirated software could be moved.

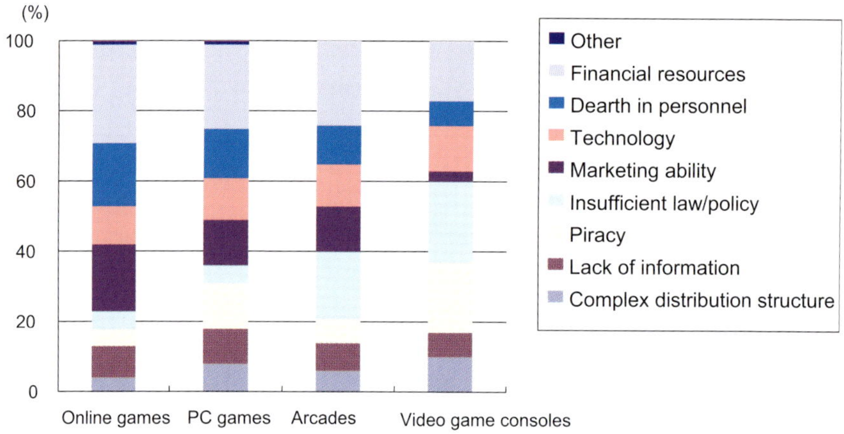

Figure 5.2 Factors challenging game companies in Korea

Source: Korea Game Annual Report 2002 Edition

Figure 5.2 shows results of a survey conducted in February 2002. 310 Korean game development and distribution firms participated in the survey, which asked firms to highlight the primary difficulties facing them in the game market. The responses show that there are vast differences in the occurrence of piracy between the different types of game.

20% of companies related to console gaming claimed to have suffered from piracy, a factor second only to insufficient policy in its frequency of being reported as an issue. PC game developers reported a lower but still substantial incidence of piracy at 13%, 3rd being financial resources and a lack of qualified personnel at 24% and 14%, respectively. Only 7% of arcade game developers viewed piracy as a threat to their business, and almost no online game developers felt piracy was something they needed to be concerned with.

The emergency of online gaming provided developers with a business model which thwarted the threat of piracy with a single blow. Because online games are provided via connection to a server run by a game developer or an ISP rather than being sold in a package, piracy is precluded.

The lack of piracy caused many Korean game developers to switch their efforts from offline to online game development, which accelerated

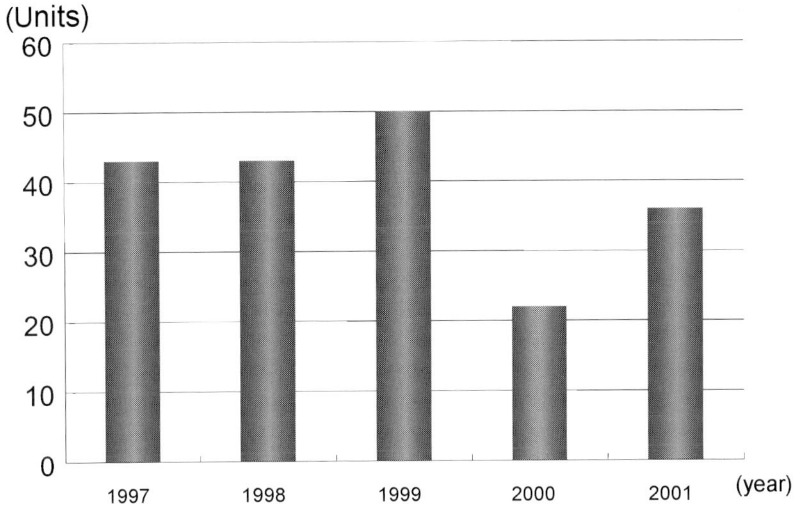

Figure 5.3 Release trend of PC games in Korea

Source: Korea Game Annual Report 2002 Edition

the growth of the industry. Figure 5.3 shows that the release of PC games by Korean developers was accelerating until the year 1999, when 50 titles were released. But there is a sudden drop in 2000 when only 22 titles were released, followed by 36 in 2001. Of the games released in 2001, 15 were children's games, thus the actual number of titles was only 21.

Furthermore, by examining the 325 games submitted to be included in the 2001 Outstanding Game Dictionary, the results are similar. The ratio of PC games fell from 42% to 25% in 2000, while online games surged to 43% of submitted titles from just 11% in the previous year. This highlights a massive shift of PC game developers to online games.

3. Supporting infrastructure

New industries are fragile, and require the existence of infrastructure to lend support and supplement the product in order to thrive (Abernathy and Utterback, 1983). Gas stations, for example, contributed significantly to the spread of automobiles in America in the early 20th century. Before Ford introduced the Model T, automobiles powered by electricity and steam competed with those powered by gasoline to be the dominant automobile design. The readily available supply of fuel via the gas station infrastructure enabled the triumph of the gasoline-powered automobile (Abernathy, 1978; Abernathy *et al.*, 1983).

The Korean online game industry's growth was driven by the existence of a strong infrastructure, similar to that of the gas station and the automobile. Internet cafés, which enabled mass distribution of games, are an example of this infrastructure.

The infrastructure which contributed to the rise of the Korean online game industry possessed a few defining attributes. First, none of the elements were deliberately planned or controlled by a specific entity, such as the government. As discussed in detail later, each factor was driven by different entities and market forces and emerged at different times. For example, Internet cafés and online game developers were two separate fields with no real relation to each other, but the introduction of Lineage aligned their interests and created a symbiotic relationship. The same applies to the spread of ADSL broadband Internet access, mobile small-amount billing systems, and the commercialization of Avatar items.

This stands in contrast to the infrastructure in other developed nations such as the U.S. and Japan. No supporting infrastructure had emerged to back the expansion process of online games in these countries. In Japan, minors are precluded from paying for online gaming as payment via cell phones is not an option. Minors, being unable to pay, were prevented from playing online games, which prevented the spread of such games. In the U.S., credit cards are the payment method of choice, which minors are not able to use without parental permission. While the infrastructure in the U.S. and Japan impedes the spread of online gaming, the infrastructure of Korea evolved to support it, and thus online gaming has spread faster and farther in Korea than elsewhere.

Figure 5.4 further highlights the infrastructure elements supporting the Korean online game industry. As mentioned previously, three

Table 5.1 Emergence of each factor

	Infrastructural factors
1996	Nexon launches Kingdom of the Wind
1997	The Korean government makes a transition to fostering communications competition
	Hanaro Telecom (the key player in the spread of ADSL) founded (October)
1998	The beginning of the spread of ADSL
	Starcraft popularized
	Internet cafés begin to spread rapidly
	NCsoft launches Lineage (September)
2000	Small-amount billing method via mobile phone introduced (July)
	Emergences of the Avatar business (November)

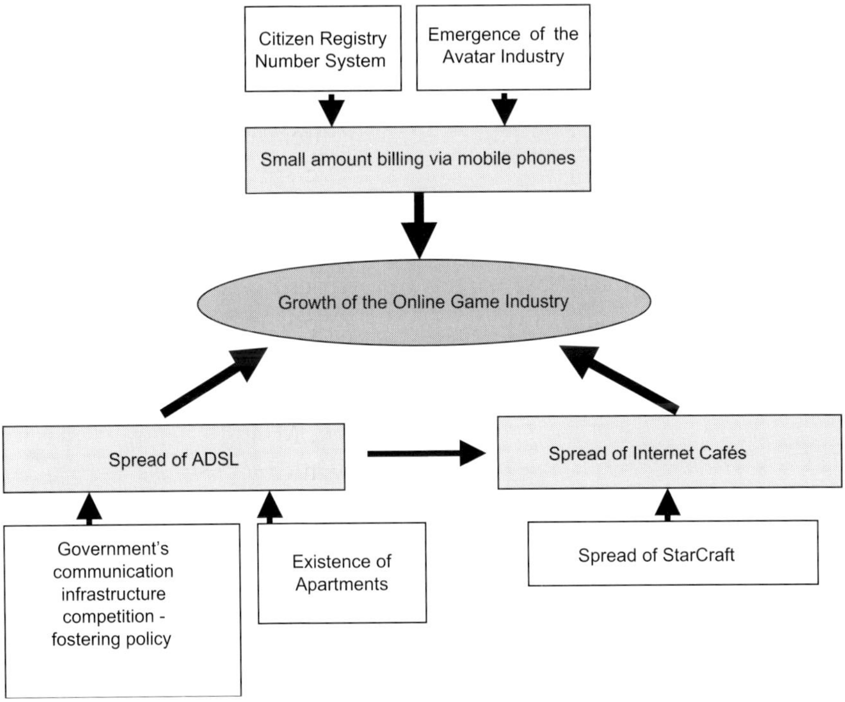

Figure 5.4 Infrastructure promoting the formation of the Korean online game industry

major factors contributing to the spread of online gaming were the prevalence of Internet cafés, the spread of ADSL service, and the rise of mobile small-amount billing systems. And yet, as the figure shows, each of these factors arose not due to online gaming but as a result of other market conditions.

1) *Spread of Internet cafés*

The "PC Bang (room)" is a business model created in Korea. PC Bangs are different from game centers developed in Japan and the Internet cafés of Europe and the U.S. These Korean cafés are equipped with anywhere from 10 to 100 or more computer terminals which allow many people to simultaneously access the Internet and play online games. Most jobs requiring a PC can be performed at a Korean Internet café, such as word processing, printing, and scanning. Internet cafés first appeared in Korea in 1995, but until early 1998 their number held steady at around 100 nationally.

Internet cafés in Korea are often called "game rooms", as the players of various console games account for more than 80% of their

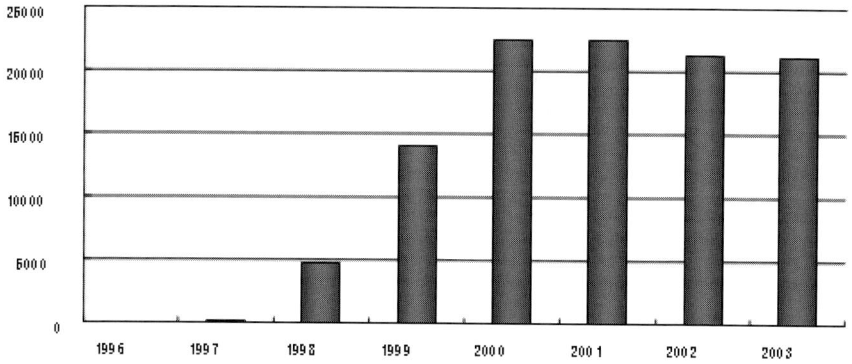

Figure 5.5 Change in the number of Internet cafés

Source: Korea National Statistical Office Official Web Site, 2003 Industry Size — Game Rooms[2]

[2] Korea National Statistical Office documents referred to "PC Bang" Internet cafés as "game rooms."

Figure 5.6 Interior of an Internet café in the U.S.

patrons. As online games became prevalent in Korea, the Internet café industry boomed by offering fast connections at low prices. By late 1998 their number had surged to 4,700, and by the turn of the millennium there would be over 20,000 Internet cafés in Korea.

Internet cafés became a cultural venue for teenagers and college students, and promoted the formation of intra-game communities which helped cross-market titles. While American or Japanese gamers played alone or with a handful of friends at their homes, Korean youths converged at Internet cafés to play games. They formed teams for competitive play and made friends with other gamers. Such communities are the forerunners of the large online cyber communities where thousands of players come together to play games, as in Lineage.

Internet cafés in the U.S., Japan and France are very different from those in Korea. Korean cafés are viewed as a venue for culture and leisure, a popular spot to gather and meet friends for an afternoon or evening of enjoyment. Rather than being a place to play games, Internet cafés in the U.S., for example, are seen as places for simple Web surfing and refreshment. This difference in the way Internet cafés are viewed and utilized in different cultures explains the faster growth of the online game industry in Korea compared to other nations.

Figure 5.7 Interior of a Korean "PC Bang" Internet café

The role of Internet cafés in the teaching of PC skills

Internet cafés in Korea have assisted with the training of the populace in computer skills. Online games, especially more advanced ones, require quick and accurate use of the keyboard and the mouse to succeed. The expansion of Internet cafés not only offered the online game industry a convenient and widespread way to market their games but also helped to familiarize new users with keyboard/mouse and Internet usage.

The commercial success of online games requires a populace familiar with basic computer skills. Typing skills, for example, are important for communicating with the other users of a game and developing relationships with other players. This is why Namco created "Led Zones" in Japan, which are Internet cafés that also provide training to users. Namco's intent was to raise users' computer skills and make their games accessible to new players.

The securing of stable financial resources by game developers

Another important function of Internet cafés was to secure a stable revenue flow for game developers. Developers charge Internet cafés a set monthly fee to access their games. This is known as a paid IP system, as the café is assigned IP addresses that are able to connect to a game's servers with no additional cost. Via this system, users at the Internet café can access a game for no additional cost, and the monthly fees provide a predictable stream of income for developers. The stability is important, online game industry revenues normally can fluctuate greatly from month to month, making it difficult to plan for expenses.

> "Allowing kids at an Internet café to play games for free is what the paid IP system enables. Kingdom of the Wind started it, and as rumors circulated that Kingdom of the Wind was free at certain locations, people swarmed into the Magic Plaza and Icon, Internet cafés near Yonsei University. The most popular by far were a few spots in Shilim-dong. That was the beginning of the paid IPs. At first, I personally accepted paid IPs because my business required it, but NCsoft officially set a price for the Internet cafés. That's why we refer to NCsoft as having started the systematization of paid IPs." (interview with an Internet café owner)

As the café owner interviewed above states, in the early stages of online gaming, Internet cafés and online game developers established a paid IP system out of mutual need. Players of Kingdom of the Wind, for example, paid a fee of approximately 2 cents per minute when connecting via a PC communication network, but at an Internet café they could play for free. The café paid a monthly fee of roughly $200 for unlimited access to the game.

Later, NCsoft would make heavy use of this arrangement to promote Lineage, and this would cement the paid IP model as one of the primary revenues sources for online gaming.

How the popularity of Starcraft promoted the spread of Internet cafés

Starcraft, a strategy simulation game from Blizzard Entertainment, has been one of the most popular games of all time and was especially successful in Korea. Starcraft is a multiplayer game in which players choose one of three factions and wage war against one another. One of the keys to its success is Blizzard's offering of a free multiplayer service called "Battle.net". Even before Battle.net was launched, gamers were playing Starcraft via paid services such as Khan, Kail, and Case's Ladder. When Battle.net was released and provided multiplayer service for free, users flocked to it. Battle.net also has a unique ladder ranking system which is updated every three months, and causes gamers to invest a great deal of time playing Starcraft in order to retain their rankings.

Starcraft was enormously popular in Korea. By October 1999, 16 months after the Korean launch, 1 million copies had been sold. By 2004, over 3 million had been sold and Starcraft was the number-one selling game in Korea for 7 years. Korea accounted for one-third of all international sales of Starcraft, and Korea ranked first in the world for Battle.net usage. In April 1999, out of the top 100 ranked players in the world, 61 of them were Koreans.

Internet cafés were crucial to Starcraft's growth, and vice versa. Internet cafés had begun to spread in 1998, the same year Starcraft was launched, and it was the perfect game to spur their growth. Starcraft had a low network load compared with contemporary titles and it could be played on a computer with a 133 Mhz Pentium chip. It was the perfect game for Internet cafés to offer.

As Starcraft grew in popularity, Internet cafés experienced rapid growth. Most Korean residences used dial-up connections, while Internet cafés offered high-speed broadband connections. Users on dial-up connections had slow speeds and couldn't fully enjoy Starcraft. Players flooded Internet cafés to play Starcraft on Battle.net, and 1998 ignited explosive growth in the Internet café industry.

2) *Spread of ADSL*

The second factor contributing to the rapid growth of the online game industry in Korea was the spread of broadband Internet service. Asymmetric Digital Subscriber Line technology, or ADSL, was quickly adopted in Korea after its invention. Internet cafés hurried to get ADSL connections, which enabled them to compete effectively against PC communication firms that allowed direct Internet connections in residences, albeit at much slower speeds.

Two important factors enabled the spread of ADSL in Korea. The first was the government opening up competition within the telecommunications industry. Prior to 1997, KT (Korea Telecom) held a monopoly in telecommunications, which was revoked with a new government policy that encouraged competition. After this policy was enacted, Hanaro Telecom was founded in October 1997. Hanaro intended to compete against KT for telephone service, but quickly realized it was facing stiff competition. Hanaro's director Ko Jin-woong recounts their dilemma.

> "KT had under its belt more than 100 years of history with telephones. Upon actually launching the business, we began to realize we couldn't compete with KT for telephones. That's why we avidly sought a sector where we could prevail — that was ADSL. At the time, the Internet was being accessed via dial-up modems, but one could not go online and talk on the phone at the same time. That is why we came to focus on high-speed Internet access while looking for a way to enable simultaneous telephone usage and Web surfing. In April 1998 we started a set-fee ADSL service in 4 cities including Seoul and Pusan."

Hanaro Telecom's offered fast, affordable ADSL service to residential subscribers. They had two plans, a 2 Mbps connection for roughly $28 per month which accounted for 70% of their subscribers, and an 8 Mbps connection for $38 per month. These low prices allowed for the rapid spread of ADSL throughout Korea.

Another factor working in Hanaro's favor was the prevalence of apartment living in Korea. Currently over 50% of Seoul's population live in apartments, large complexes built to house thousands or even tens of thousands. ADSL connections require the phone company to connect optical cable to an apartment complex's basement maintenance facility, which then passes through the MDF (Main Distribution Frame) of the complex before splitting off to separate residences via copper phone wire. The phone company is responsible for the costs of laying cable up to the basement maintenance facility, thus they can install connections that reach many families with one construction project. They avoid the "last one mile problem" that telecom firms face when offering service to detached homes, and are able to install ADSL at a low cost.[3]

The easy availability of high-speed Internet connections brought about changes in the market. Internet cafés, which relied on fast and stable connections, grew rapidly by adopting the new low-cost, high-speed ADSL service. The broadband network also saw rapid acceptance by residential users, and gamers who had previously played online games at Internet cafés were now able to play from the comfort of their own homes. As Korea was one of the earliest adopters of broadband technology, there was not a lot of content available on the Internet that took advantage of high-speed connections.

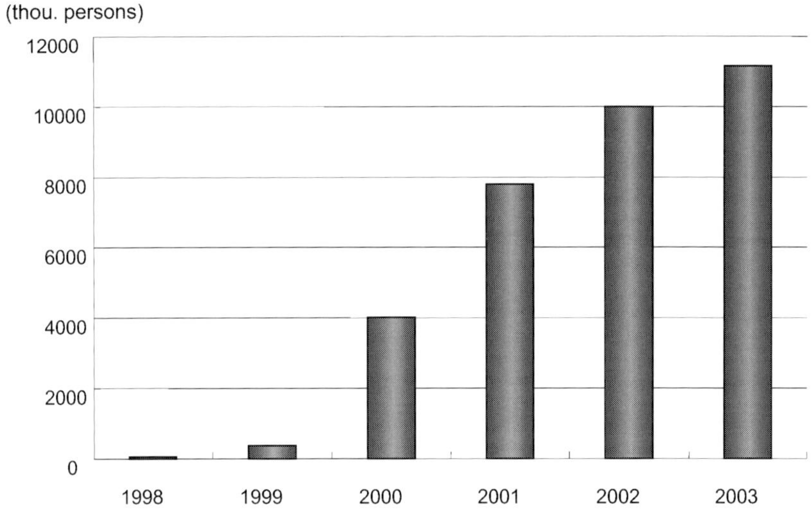

Figure 5.8 Change in the number of residences registered for high-speed internet access

Source: Korea National Statistical Office 2003 Household Income — Number of Households Signed On for High-Speed Internet Access

[3] The "last one mile problem" is the final leg of delivering connectivity from a communications provider to a customer. It is typically seen as an expensive challenge because "fanning out" wires and cables is a considerable physical undertaking.

Online gaming was about the only avenue where a high-speed connection could truly shine; thus, the spread of ADSL encouraged more Koreans to play online games.

3) *Mobile small-amount billing system*

In Korea, cell phones were designed such that they could be used to pay for small services, such as buying a soda from a vending machine or paying for an online game. This is known as the small-amount billing system, and the fees are simply added to the monthly phone bill.

Korean online game developers offer a wide variety of payment methods, including prepaid game cards, credit cards, and direct deposit. Most gamers, however, opt to pay via their mobile phones. Such a simple and convenient method of payment makes it easy for users to play online games.

Mobile phone companies levy a service charge ranging from 12.5% to 15%, much higher than the fees charged by credit card companies, which range around 3%. But because users overwhelmingly prefer to pay via their mobile phones, game developers have no choice but to offer small-amount billing as a payment option.

Anyone who owns a mobile phone can begin playing a game immediately after paying via this method, as the payment is instantaneous. This is particularly convenient for teenaged gamers who have mobile phones but no credit cards. Of the payments processed by game developers NCsoft and JC Entertainment, 70% and 80% of them are made via mobile phone, respectively.

The process of payment via mobile phone, illustrated in Figure 5.9, is as follows.[4]

1) The user provides the content provider with his Citizen Registry Number and mobile phone number.
2) The content provider transmits and confirms the information via a PG (Payment Gate) firm, which acts as an intermediary.

[4] Compiled from an interview with representatives from Info-Hub and Mobilians, a large PG firm offering cellphone small-amount billing.

3) The PG firm transmits and confirms the information via the communication carrier server (mobile phone provider).
4) If the information is correct, an authorization number is sent to the user's mobile phone.
5) The user enters the authorization number into the computer, and gains access to the purchased content.
6) Fees are deducted by the intermediary firms. The communications carrier receives the payment and deducts a service charge of 2.5% to 5%, and then passes the payment on to the PG firm. The PG firm subtracts their charge of 10%, and the content provider receives from 85% to 87.5% of the payment amount.

In Figure 5.9, the numbers in the squares refer to the order in the payment process. Dark lines denote the flow of information, and light lines the flow of fees.

Two factors contributed to the popularity of mobile phone small-billing systems. The first is the existence of the Citizen Registry Number, and the second is the rise of the Avatar industry.

The Citizen Registry Number system registers all Korean citizens over the age of 18 in a government database. This number can be used to identify a specific citizen, much like a U.S. Social Security

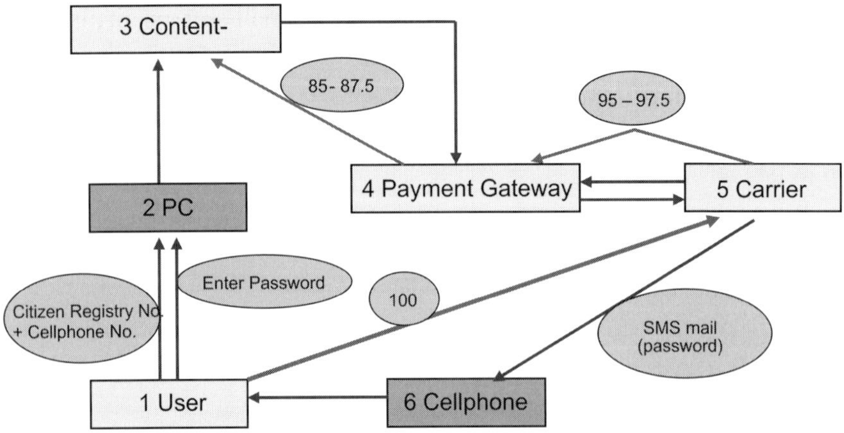

Figure 5.9 Mobile small-amount billing system

Number or driver's license number. As PG firms researched the viability of mobile phone billing systems, they initially tried to offer a membership-based service where users signed up and were provided with an authorization number. This method was not successful as people were reluctant to provide additional personal information to identify themselves. The PG firms hit upon the idea of using the Citizen Registry Number, which is already supplied to mobile phone providers when an account is opened, and thus there is no need for additional private information to be provided to the PG firm. All a user needs to sign up for small-amount billing is a mobile phone number.

The second factor is the emergence of the Avatar industry. The word "Avatar" describes a physical incarnation of a Hindu deity, but in computer terminology is a graphical representation of a user while in a game. As shown in Figure 5.10, users can select physical attributes of their characters on an Avatar business site, and then incorporate all kinds of additional features like clothing, shoes, and wigs to make the Avatar look like whoever they desire. These additional features are known as "items", and users pay for them to enhance their avatar's appearance in-game via small-amount billing.

In the early stages of mobile phone billing systems, the reaction from Internet content providers was somewhat dubious. The system was new, and the PG firms running it were small startup firms without a commercial history. Also, most content providers had not gone commercial at this time. Most content providers tried to draw in as many users as possible by providing content for free and then advertising to them. The number of registered users and the frequency with which they visited the site were important thus content providers offered their content for free to keep the largest possible pool of users.

As the dot-com advertising boom collapsed, content providers were forced to go commercial, and many of them began looking into relationships with mobile phone billing providers. Say Club, Korea's largest chat site which launched commercially in November of 2000, was one of the first adopters of mobile phone billing. Lee Kang-seop

Figure 5.10 An example of Avatar items

Source: <http://www.sayclub.com>

of Mobilians describes the relationship between Say Club and mobile phone billing below.

> "Sales in our first month in business were just $300. We deemed that to be too low for a firm with 20 employees. As we were struggling, Say Club surfaced. In this chat site male/female characters

designated by users engage in conversation. The vast majority of users are young men and women hence they all want to make their characters more appealing. The Avatar business involves purchasing items to embellish characters. For example, users adorn their characters with a $0.35 pair of shorts, $0.15 pair of socks or 0.95 suit. Say Club sold bouquets, suits and chocolate boxes for men for $5 each on Valentine's Day and were pleasantly surprised with the explosive response. Mobile phone billing was an ideal payment method for such purchases."

The price of any given Avatar item was quite low, thus paying via credit cards would be unwieldy due to the relatively high base service charge per transaction. Payment via mobile phones, which was just coming in to the market, was ideal for Avatar items. Via a website such as Say Club's, users can conveniently purchase an item and receive it in their game account immediately. The Avatar business provided much revenue for early PG firms, and helped contribute to the rise of small-amount billing.

4. The support of governmental policy

Since the economic growth plan instituted in the 1960's, the Korean economy has adhered to a formula of the government leading and civilian industry participating. From chemicals to electronics, Korean industry has historically been kick-started by governmental work to create a framework and provide incentives for businesses to start up in selected fields.

The online game industry was the first to break that mold. The core of the field was formed without government aid, and even occasionally in the face of government suppression under the auspice of protecting teenagers from inappropriate online content. As online gaming grew and became viewed as an actual industry, the government began to institute policies fostering it and encouraging its growth. But the growth had begun independent of government involvement, and this marked the first civilian-led industry in Korea.

Table 5.2 State of government's game industry-related policies

Implemented policy	Year implemented
Music, film and games legally institutionalized/enforced	1999
Korea Game Development and Promotion Institute founded	1999
Korea Media Rating Board classification on games strengthened	2001
Game industry firms recommended for military service	2000
Game industry venture capital firms certification evaluation implemented	2000
National skills qualification system enforced in the game industry	2002

Source: 2004 Korea Game Annual Report excerpt

As shown in Table 5.2, the government did institute policies to encourage growth in online gaming once the industry was solidly established. Some of these policies proved very useful in propelling the industry's rapid growth. Two especially important policies are: the communication infrastructure competition policy implemented by the Ministry for Information and Communication, and the policy of adapted military service for game developers implemented by the Ministry of Culture and Tourism.

As discussed previously, there was no competition in the Korean telecommunications industry up until the 1990's. When the industry opened up and other firms entered the marketplace as ADSL became popular, the competition brought down prices and made high-speed Internet service widely available. This triggered the rise of Internet cafés and residential high-speed communication networks, which were important factors in the growth of online gaming.

Adapted military service, which provided incentives to engineers and programmers to work in the game industry, helped game developers in recruiting a skilled workforce. Most Internet startups had a difficult time recruiting qualified employees, as large corporations worked hard to recruit skilled IT personnel. This incentive policy was critical in supplying the workforce needed to grow the online gaming industry.

(1) *High-speed communication network policy (Ministry of Information and Communication)*

The building of high-speed communication networks was essential for the growth of online gaming. When the Ministry of Information and Communication eased regulations in the telecommunications market, the resulting spread of high-speed Internet networks built a foundation upon which online gaming could grow.

In the early 1990's, the Ministry of Information and Communication began studying methods by which to encourage competition in the telecommunications industry. The Ministry implemented a plan of corporate restructuring, known as the "High-Speed IT Foundation Formation Implementation Plan", which was divided into 3 stages to be implemented between 1995 and 2015.

In 1993 and 1994, the government realized that a solid communications infrastructure was essential for Korea's growth. They formed the High-Speed Communication Network Formation Committee to research the effort, which was made up of 49 representatives from 6 government agencies and 6 communication-related firms. The plan was modeled after the information highway plans of the U.S. and Japan. The basic plan was to ease market regulations via enacting policies to ensure fair competition and promote entry into the market. They also took measures to break up monopolies and prevent dominant companies from abusing their position and engaging in unfair competition.

At the time, the communications industry was classified into the companies which owned the communication networks (period communication firms) and companies leasing access to the networks (additional communication firms). These were reclassified into monopolies and competitors. Regulation on monopolies tightened while it was loosened on competitors to encourage growth. Seo Hong-seok of the Ministry of Information and Communication states, "The important thing to keep in mind is injecting competition into the communications industry, going from KT's monopoly to the emergence of Hanaro and Dacom. Competition has made an expansion of the infrastructure possible."

Most countries adopted local loop unbundling in the early stages of market formation, but it was adopted in Korea when the market was already growing after equipment competition was already adopted in the early stages. This policy caused firms to engage in competition based on equipment and technology rather than service, which favored consumers as service prices remained low.

Eliminating entry regulations for firms seeking to provide high-speed Internet access enabled rapid market growth as many companies entered the field. The government realized that minimal regulation would be most effective in promoting competition in the market's early stages, and they were right.

These policies encouraged the formation of Hanaro Telecom in 1997, but they began as a telephone provider competing with KT (Korea Telecom) and ran into early difficulties as discussed previously. Hanaro, along with KT and Dacom, began looking into the viability of providing broadband Internet service via telephone lines instead of cable.

In June 1998, Thrunet began to offer Internet access via cable modems for a relatively low cost of $40 per month. This encouraged Hanaro and KT to speed their development process, and Hanaro introduced their ADSL service in April of 1999 with KT following in June. Three major players were now competing in the high-speed Internet service market, which kept prices low for consumers. The availability of high-speed access aided the rapid spread of Internet cafés, and Korea became one of the first countries to widely adopt broadband service.

(2) *Adapted military service system (Ministry of Culture and Tourism)*

Korea has a policy of mandatory military service; however, citizens can fulfill their military obligations by going to work for certain firms under the adapted military service program. This system was implemented in March 1973 to foster economic growth by providing key industries with skilled workers. Draftees can work for between 26 and 34 months in an eligible field and receive full credit for their military service. In 2000, the Ministry of Culture and Tourism began recommending game developers for the adapted military service program.

Table 5.3 Game industry adapted military service designation status

	Classification	New		Existing		Total	
2000	Number of firms applied/ recommended	101 firms	65%	18 firms	100%	119 firms	70%
	Number of firms selected (Number allocated)	66 firms (102 individuals)		18 firms (55 individuals)		84 firms (157 individuals)	
2001	Number of firms applied (recommended)	84 firms (68 firms)	58%	79 firms	98%	163 firms (147 firms)	77%
	Number of firms selected (Number allocated)	49 firms (49 individuals)		78 firms (165 individuals)		127 firms (214 individuals)	
2002	Number of firms applied/ recommended	20 firms	20%	97 firms	47%	117 firms	43%
	Number of firms selected (Number allocated)	4 firms (4 individuals)		46 firms (47 individuals)		50 firms (51 individuals)	
2003	Number of firms applied/ recommended	11 firms	27%	78 firms	24%	89 firms	25%
	Number of firms selected	3 firms		18 firms		21 firms	
2004	Number of firms applied/ recommended	11 firms		56 firms		67 firms	

Source: Military Manpower Administration Internal Data

Prior to 2000, it was difficult to qualify for adapted military service in the game industry as game developers were somewhat awkwardly classified as part of the information-processing industry. Game developers were added to the Standard Industry Classification Code in 2000, and the government began designating firms eligible for adapted military service in conjunction with the Military Manpower Administration. This helped to provide game developers with a pool of skilled workers, as in Table 5.3 highlighted.

Most employees who entered the gaming industry via adapted military service were assigned jobs in game production and development, the most important tasks for a game developer. This was a great boon to game developers, as previously they had a difficult time competing with large corporations to attract employees. Larger companies could offer higher salaries and better benefits, and adapted military service helped to counteract that and attract employees to game developers. A vast majority of workers in the game development industry began via adapted military service, and the program played a key role in the success of the industry.

The list of game developers who participated in the adapted military service system is long and prestigious. Kim Taek-jin, CEO of NCsoft, served at Hyundai Electronics. Kim Bum-soo (president of Web-game developer NHN), Park-Jin-hwan (president of Neowiz), Kim Jung-ju (president of Nexon) and Yoon Seok-ho (president of CCR) all started their careers in the game industry via adapted military service. Kim-Jae-hyung, the president of Daum Game, served for 5 years at the LG Institute of Technology and Daum Communications. Kim Hak-kyu, creator of the online game "Ragnarok", served at Nahana Computers in Seoul. This is just a short list to highlight the importance of adapted military service to the growth of Korean online gaming.

Being designated as eligible for adapted military service also boosted a company's corporate image, as it demonstrated a trust in the company by the government. Kang Sam-seok, president of Masangsoft, states the following.

> "Mid-sized firms like us cannot attract skilled employees. The adapted military service system purveys mid-sized game developers

> with a stable personnel supply. My company does not discriminate between ordinary employees and those serving military service, and everyone's treated equally. If we don't treat them equitably, problems arise in corporate unity, and the individual serving his military service can harbor complaints. Also, being designated as a firm eligible for adapted military service produces the effect of creating public trust; it instills the image of us as having been preliminarily recognized by the Military Service Administration and Ministry of Culture and Tourism. In such a case we will make a positive impression on customers and investors."

The adapted military service program is viewed by young Koreans as a great opportunity. Rather than wasting one's talent for 2 years in the military, eligible workers could build valuable experience and skills. Game development requires high-level programming skills, which must be constantly kept up to date. Spending 2 years in the military is a long time in which those skills will atrophy, so interested workers are eager to join the industry as quickly as possible.

Chapter 6

Global Development and Marketing Strategies of Korean Online Games

Korean online games have not only been successful in the domestic market, but are beginning to break into international markets as well. In 2007 Korean online games were offered in many markets, with international revenue totaling approximately $800 million. Japan is currently the largest market for Korean games, accounting for 32.4% of exports, China is second with 24.6% of export revenue and Southeast Asia with 7.2%, the U.S. with 19.2%, and Europe with 5.9%. Korean online games boast a total market share of 36.5% in the global online gaming market.[1]

Korean online games have so far seen the greatest success in China, Japan, and Southeast Asia, with a roughly 50% market share in all three regions. Titles such as Cart Rider, BnB, Legend of Mir 2, and Ragnarok have been instrumental in leading the global charge of Korean game developers. Of the top 10 online games in China in 2008, four of them were Korean titles.

The Japanese online game market has been dominated by Lineage 2 and Ragnarok, with Ragnarok's December 2002 launch drawing 100,000 registered users on the first day. Currently, Ragnarok has over 1 million Japanese players, and has led the growth of the Japanese market. GungHo Online Entertainment, the Japanese game publisher that services Ragnarok in Japan, was listed on the Osaka

---

[1] Online Game White Paper 2007.

Table 6.1 Top 10 Chinese online game — February 2008

Ranking	Game title	Developer
1	WOW (World of Warcraft)	Blizzard
2	KartRider	NEXON (Korea)
3	Crazy Arcade BNB	NEXON (Korea)
4	Zhuxian	Perfect World
5	Audition	T3 ENTERTAINMENT (Korea)
6	Fantasy Westward Journey	NetEase
7	FreeStyle	JC ENTERTAINMENT (Korea)
8	Tian Long Ba Bu	SOHU
9	Westward Journey 3	NetEase
10	CQ Online	Kingsoft

Source: <www.popsoft.com.cn>

Heracles stock exchange in March 2005 and quickly attained a market capitalization of approximately $3 billion. GungHo's market cap is comparable to that of major Japanese game firms such as Square Enix and Koei, demonstrating investors' confidence in the growth of online gaming in Japan.

1. Factors affecting global online gaming growth

Experience has shown that there are three key factors that influence how online gaming will develop in a given market.

The first of these factors is the status of the domestic console game market prior to the arrival of online gaming. An entrenched console and PC gaming industry will lead to slower acceptance of online gaming. In Japan, for example, there is a strong offline gaming industry, led by experienced game developers including Nintendo, Sega, and Sony. Japanese gamers have exhibited strong brand loyalty, and the growth of Korean games in Japan has been slower than in China. It is still possible to successfully market online games in a country where there is an established offline game industry, but acceptance tends to be at a slower pace than in a country where no offline game industry exists to serve as a competitor.

The second factor is the level of computer literacy of the populace. In Korea, the extreme popularity of PC games such as Starcraft trained players in the use of PCs, including skills such as typing that are valuable for online games. This familiarity with computer controls leads to a quicker embrace of online games, as games seem more accessible to new players.

The third factor is the prevalence of piracy in the offline game market. As was previously discussed, piracy was rampant in Korea prior to the rise of online gaming. This led game developers to embrace online gaming as a way to combat piracy and bring in more revenue from their games, and this trend is likely to continue in other markets as well. If piracy is a threat, game developers will be more likely to migrate to making online games for the greater security they offer.

Consideration of these factors leads to the conclusion that online gaming will spread rapidly in China, Taiwan and Southeast Asia. These countries share the traits of not having an entrenched offline game industry, having a reasonable base of computer literacy, and having issues of software piracy. Conversely, online gaming will likely see slower growth in the U.S., Japan, and Europe. The conclusion to be drawn is that there is opportunity for online games in all of these regions, but the marketing approach must be appropriate to the country in which a game is being released.

Another important factor is that foreign expansion of online games is far different from that of PC and console games. Exporting offline games is similar to exporting industrial products. The game gets rewritten into the local language and sold, but once the sale is made, no additional work is required on the part of the developer.

Exporting online games to new markets is far more complex. The game must be localized to the domestic market, but once the game is published and users are accessing it, the work is only just beginning. The game developer must offer customer service in the local language, combat the activities of hackers, and offer regular updates of game content. This requires a strong knowledge of the local player base, so that updates can reflect their desires for the game and encourage them to keep playing. It is not uncommon for a game that is published in many different markets to have different patches in each market, catering to the tastes of disparate cultures of gamers.

Running an online game is a difficult enough prospect in the game developer's domestic market. Servicing an online game in a different culture presents a terrific challenge, both in technical issues and in customer service. China provides stark examples of this, as there are many differences to be accounted for in the various regions of the country. There are vast differences from region to region in Internet infrastructure, for example. When a game developer plans an event or a marketing campaign in China, these regional attributes must be taken into account. Chinese Regional CEO, Chu of Asia Game has the following to say on this issue.

> "Users are primarily concentrated in the northeast in a game 'One Hundred Year'. As northeastern users hold almost no gatherings involving singing or dancing we attract users via offline events, activity in Internet cafés or contests. But such tactics will fall through in Guangzhou where Red Moon users are concentrated. As numerous gatherings involving singing or dancing take place in Guangzhou we gave up on adults and hold events primarily targeting schools or students. For example, we hold an event in conjunction with the student government at a college, or at an Internet café near a school. And because students don't have disposable income we can secure users by offering items or game money to them if they refer a friend. There are many regions in China. For example, the east is affluent, the west is poor, people in the north lead simple lives and people in the south lead fast-paced lives. That is why reception to popular games differs by region."

Game developers looking to expand internationally must consider three broad issues. Arriving at an appropriate combination of these factors is key to the success of their foreign expansion efforts.

1) Server operation: Global or local servers
2) Publishing method: Direct or outsourced publishing
3) Distribution method: Package sales or online downloads

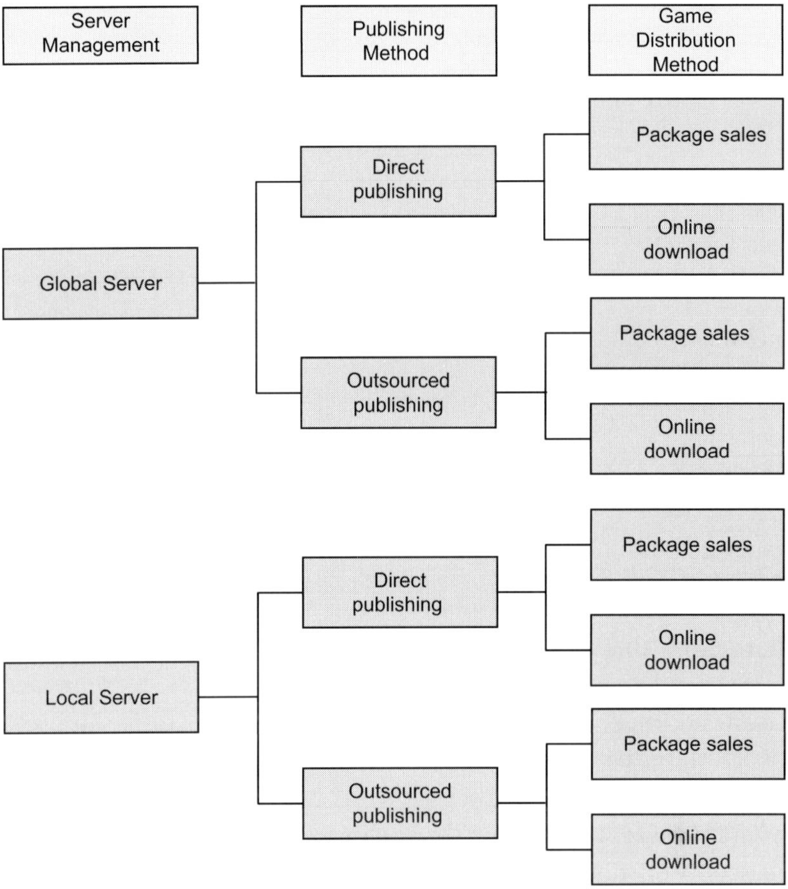

Figure 6.1 Global service method of online games

1) *Server operations*

When a developer is looking to offer their online game in a new market, they must consider whether to use global or local servers. A global server can be accessed from anywhere and thus combines players of different cultures on one server, while a local server is specific to the new market and only caters to players from that area.

There are advantages and disadvantages to each choice of server type. Using global servers means that the servers can be hosted at a

Table 6.2 Server operations method

	Global server	Local server
Format	Server can be accessed anywhere in the world	Server can only be accessed in a limited area
Server operations method	Central management	Dispersed management
Maintenance costs	Low	High
Billing system	Central management	Dispersed management
Customer DB	Central management	Dispersed management
Degree of policy-sharing within the organization	High	Low
Response to local customers	Slow	Quick

central location and directly managed by the game developer. This is generally cheaper, as economies of scale occur. Using global servers also ensures that policy changes and content patches are applied smoothly and simultaneously to all markets. However, customer service can be more difficult, both because of any time differences between the player and the support staff and any cultural or language barriers that inhibit communication.

Local servers require the developer to either open a new office of their own in the new market, or to partner with a local company to run the game servers and customer support department. This is generally a more expensive option, and can also lead to delays on policy implementation and the application of content patches for the game. The advantage of using local servers is that customer support tends to be better, as a local support staff is more likely to communicate well with their users in a timely fashion.

2) *Publishing method*

Publishing methods are comprised of direct and outsourced publishing. Direct publishing entails the game developer publishing the game themselves in the new market, handling localization of the new

Table 6.3 Publishing method

	Direct publishing	Outsourced publishing
Format	Performs distribution and marketing	Distribution and marketing outsourced
Operating entity	Game firm	Publisher of the region
Early penetration of the market	Difficult	Easier
Securing customers early on	Difficult	Easier
Marketing	Centralized	Dispersed
Game brand value	Low	High

game version and all marketing efforts. Most online game developers are relatively unknown outside of their domestic market, however, which means they have to start from the ground up building their brand in a new country.

Outsourced publishing involves hiring a local company to distribute, promote, and market the game in the new country. An example of this is Blizzard Entertainment hiring Chinese publisher The9 to release World of Warcraft in China. This allows the game to benefit from the established brand of the local publisher, which can lead to better market penetration and faster growth. The downside to hiring a local publisher is that the revenue for the game is split, meaning less of the profit makes its way back to the developer. Also, many such outsourced publishers handle many games at once, meaning that the publisher's marketing efforts are likely to be dispersed amongst multiple titles.

3) *Distribution method*

The third factor to be considered is the method of distribution. Developers can opt to allow their game to be downloaded via the Internet, or to sell packaged copies of the game via local retail outlets.

Package sales demand a higher initial outlay of cash on the part of the developer, as the packages must be created and shipped to their locations. Package sales allow the developer to take advantage of the existing retail network, displaying their game on shelves in stores and generating awareness of the new release. Package sales also allow for a faster infusion of revenue, as developers realize an immediate profit for each copy of the game sold. Building an online game is expensive, and developers often need this quick turnaround of sales after spending the capital to enter a new market. Finally, offering retail copies of a game circumvents any difficulties with the local Internet infrastructure that could hamper online distribution efforts.

Online downloads allow users to simply download the game, either to their home PC or at an Internet café. This generally means that users do not pay for the game immediately, and the developer must wait for new players to subscribe before they realize any revenue. Online downloads are also highly dependent on a solid Internet infrastructure being available that can support the timely download of large files. China is a good example of a market in which the Internet infrastructure is not yet robust, and game developers tend to distribute their games offline, via local Internet cafés. Choi Ki-cheol, a Wemade executive, elaborates on the difficulty of distributing online games in China.

> "In the south or deep in the inlands like Sichuansheng the infrastructure is extremely feeble because they use modems. With our

Table 6.4 Distribution method

	Packages sales	Online downloads
Format	Game CD	Program download
Product	Physical	Electronic
Distribution network	Offline	Online
CD production cost	Some	None
Distribution investment	Large	Small
Surplus inventory costs	Some	None
Affected by communication infrastructure	No effect	Some effects
Source of revenue	(CD sales) + user fees	User fees

Legend of Mir 3, which is about 3-GB large, we had a difficult time when we first began distributing it via CDs, but it has been distributed widely since then. It wouldn't fit no matter how much we compressed it. Even Internet café proprietors were reluctant on that part."

2. The Chinese market and its distinguishing characteristics

1) *Growth potential*

China has turned into a huge market for online gaming in recent years. The online game industry in China was a $788 million market in 2006, and is anticipated to reach over $1.6 billion in 2008, $2.12 billion in 2009, representing 40% annual growth.[2] The number of users for online games in China was 40 million in 2007, growing to 18 million in 2006 and 16 million in 2005. At that growth rate, 2008 should see approximately 50 million Chinese gamers playing for online games, well over half of the total number of Chinese Internet users who indulge in online gaming. Most Chinese gamers are teenagers, which affects the marketing strategies required to reach them. 16 to 30 year olds account for 87% of Chinese online gamers, and 34% of that group are ages 19 to 22.

The vast majority of online games introduced to China thus far have been MMORPGs, which means stiff competition for any new entries to that field. Casual games such as Freestyle and Audition also became prevalent in China. Game developers should note, though, that disposable income in China is generally lower than in Japan or Korea, which means business models will have to be developed that ensure new game releases are accessible and affordable to Chinese players.

China is currently one of the largest foreign market for Korean-made online games. A combination of a strong Internet café infrastructure to train users and provide high-speed access, local familiarity with Korean game styles, and rampant software piracy issues brought

[2] http://ccidconsulting.com.

on by weak copyright laws should ensure quick and consistent expansion of the market for the foreseeable future.

The Chinese government had not been unaware of the market potential for online games in its country. Having seen the potential of the market, the government has been funneling funds to domestic software firms to encourage domestic production of online games. Competition from local companies might be light so far in China, but that is not likely to remain the case for long.

Chinese game publishers have acted as outsourcing firms for foreign games, led by major publishers such as Shanda and The9. These companies are working on their own titles as well, slowly building a workforce of developers and engineers capable of creating high-quality games. Chinese firms have also looked to acquire the source codes of existing games, such as Legend of Mir, with which to build spinoffs or just use for training purposes. Some companies have also begun to look outward to improve their level of expertise. Shanda, China's largest publisher, already made a large investment to acquire a 28.95% controlling stake in Actoz Soft, a Korean developer.

2) *Risk factors*

While software piracy is in general a good sign for the growth of online gaming, in China, there is a different kind of piracy damaging some imported Korean games. Illicit servers, or servers run illegally by somebody other than the game developer, have threatened the market share of several major Korean titles. These illicit servers provide almost identical game content to a normal server, but often offer special "modifications", such as dramatically higher acquisition rates for experience points an items. Some illicit servers are free, while others charge their users a fee to play. There are currently more than 2,000 illicit Lineage 2 servers, marketed via 30 illicit server portals.

Without a strict standard for enforcing software copyrights, the central Chinese government has maintained a passive stance in dealing with this issue. When Korean firms discover illicit servers, their only recourse is alerting local law enforcement, which has generally proven ineffective (Wi, 2008).

Figure 6.2 Illicit Lineage 2 server operated in China

Source: <http://www.thegamenews.com/>

Another risk factor in the Chinese market is macro abuse. Macros are a type of hacking program for online games, and come in a wide variety of forms. Some macro programs allow users to circumvent normal game mechanics, such as removing weight restrictions on how much a character can carry. The most common macro programs are "bots", which essentially play the game automatically, allowing the character to be online and killing monsters 24 hours a day.

Heavy usage of macro programs has a powerful effect on a game's economy. If a significant number of a game's users employ macros, they will have far more money available to them than players who don't use macros. The availability of so much money in the economy leads to widespread inflation of game prices, which means that users who do not use macros have a harder time affording to purchase the items they want. In extreme cases, this hyperinflation can lead to a complete collapse of a game's economy. The figure below reflects the widespread use of macros among Chinese online gamers.

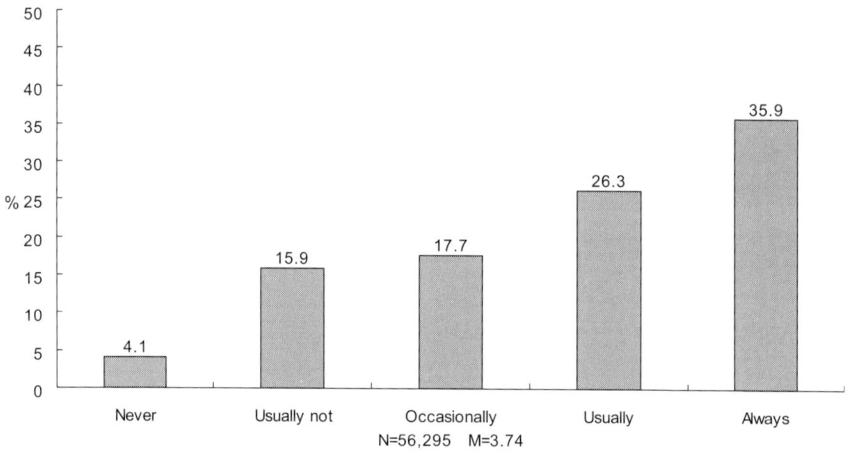

Figure 6.3 Degree of using a macro

* On a scale of 1 to 5 (1 = Never, 5 = Always)

The question "to what degree do users around you employ macros?" was posed to a survey of Chinese online gamers. 35.9% of respondents replied "always use macros" and 26.3% answered "often use macros", which suggests that well over half of all Chinese online gamers use macros to some degree or another. Choi Ki-cheol, a Wemade Entertainment executive, has the following to say regarding macro abuse.

> "Almost all users in China employ macros. They are so acclimated to macros that they won't play without them. In our game one of the most common jobs is the mage. Whereas warriors pursue and attack, mages attack while fleeing. Taking potions on time can be annoying so when the blood indicator is running low, it is replenished automatically. The weight limit is also overridden. If the limit is 100, macros which exceed that limit to 1,000 are created, then the character never dies. Blood is replenished automatically, and they are able to carry 1,000 hence they kill other characters and go after the items they drop and continue to play. The same applies to other games as well. We understand the use of macros for convenience,

but it offsets game balance. Regulations should be placed on it, but Korea is unable to take speedy measures. We should penetrate the Chinese market when we have sufficient technology to address these problems but that is not the case for most game makers."

3) *Internet cafés as online gaming strongholds*

As of 2005, approximately 100 million Chinese surf the web, and roughly 30 million of them have high-speed Internet access. Taken as a whole, though, computer ownership in China is still low, and many of them are older models, unsuited for playing online games.

Broadband Internet service is increasing, but the majority of Chinese Internet users still connect via dial-up modems. The vast disparities in economic status between the different regions of China means that most of China's Internet users reside in the eastern area of the country. Even there, though, high-speed residential access is still not common, and most online gamers play from Internet cafés. Internet cafés have emerged as the primary distribution channel for online games in the Chinese market.

There are two main types of networks used in Chinese Internet cafés. The traditional Internet café operates a large number of PCs,

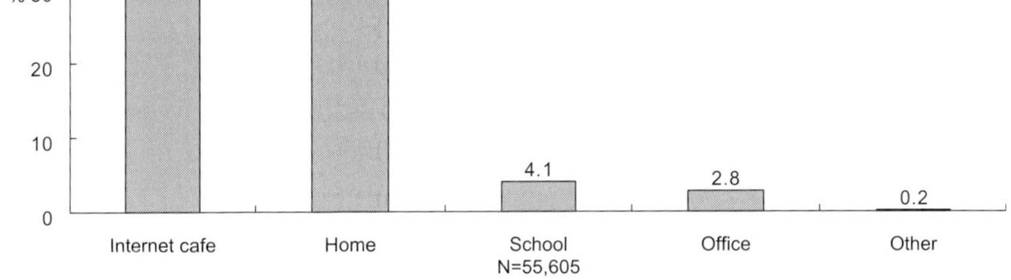

Figure 6.4 Online gameplay venues of users in China

each with their own internal hard drives on which games are installed. In these types of establishments, game developers must compete for limited available hard drive space when trying to get their games into Internet cafés. Some newer Internet cafés circumvent this issue by installing a powerful network server that hosts all of the cafés games off of its hard drive, which is generally a large network drive with far more space available than the typical single-computer hard drive.

Most Internet cafés are of the older variety, which means the first step for a game developer is convincing the café to provide their game. Kingsoft, the maker of JX Online, arrived at a novel solution to this. Kingsoft heavily advertised their game in the relatively affluent Sichuansheng region, and then negotiated with local Internet café owners, offering them very lucrative terms in exchange for making JX Online the only game available in their café.

Sichuansheng is a large region of China, with a population of approximately 100 million, and a large portion of those are college students. Sichuansheng is also a major metropolitan area, which affords it substantial influence on surrounding areas. These factors mean that a product's success in Sichuansheng is likely to spread to other areas as well.

Kingsoft's marketing strategy was brilliant. Their advertising generated substantial interest in JX Online, and their offers to Internet cafés ensured that the game was available to interested players while also removing competition. The result is that JX Online is currently the 3rd-largest online game in China in terms of simultaneous users connected.

4) *The expansion of Korean gaming in China*

The first important factor to note is that the majority of Korean online games that have been brought to China are MMORPGs. Shanda's enormously successful release of Legend of Mir 2 set the stage for the rise of online gaming in China, and many other Korean MMORPGs have followed in its footsteps.

One factor accounting for the dominance of MMORPGs in the online gaming market is the general acceptance amongst players of paying for access to an MMORPG. More casual games, such as online board games, have generally had a more difficult time transitioning to a commercial model, but pay-to-play MMORPGs have become the industry standard, and thus they tend to lead the way in new markets. As time goes by and online gaming becomes more common in China, it is likely that other types of games will spread as well, as they have in Korea.

A second interesting point is that every Korean online game launched in China thus far has been done in collaboration with a local Chinese company. Rather than establishing a satellite office of their own, Korean firms have been opting to partner with a Chinese company to release and service their games. One reason for this is that Chinese companies know the local distribution system, allowing games to get to market smoothly and be made available to the largest possible number of users. Another reason is that there are less options to pay for online games in China, as payment via credit cards and mobile phones is not yet a viable option. Users generally pay via prepaid cards purchased at Internet cafés, however, only three vendors are legally allowed to sell prepaid game cards in China. These limitations on payment make it simpler for Korean companies to partner with local Chinese firms and allow them to handle all of the billing issues.

There are three major game publishers in China. Shanda Entertainment, The9, and SINA. Shanda and The9 are both publicly traded companies on the Nasdaq stock exchange, and they have been using the funds raised by their stock sales to pursue and purchase Korean game developers..Table 6.7 discusses the three publishers in more detail.

The methods of distributing Korean online games in China have tended towards online downloads. The publisher makes the game client available online, and both Internet cafés and individual users can download the client for free. Figure 6.5 shows the commonly-used distribution methods in China, and also again highlights that the vast majority of Korean game developers are choosing to partner with local publishers to enter the Chinese market.

As Figure 6.5 demonstrates, online downloads are the primary method of distribution in China. Many publishers also offer supplemental distribution of packaged copies of the game client as well,

Table 6.5 Major publishers of Korean games

Company	Description
Shanda Entertainment (盛大)	Established in November 1999, Shanda was officially inducted into the online game business by launching Legend of Mir 2 in 2001. Legend of Mir 2 surpassed 600,000 simultaneous players connected within one year of launching, and in 2003 was ranked 1st in IDC's online game rankings. With the success of Legend of Mir 2, Shanda transformed the Our Game- and Asia Game-centered online game distribution market. The premier online game firm in the current Chinese market, Shanda was listed on the Nasdaq in May 2004.
SINA (新浪)	One of the top three Chinese portals, SINA entered the online game management market in 2002. In February 2004, SINA established IGame in a joint venture with Korean game portal NetMarble and NC-Sina with NCsoft.
The9 (九城)	The9 is an Internet startup established in August 1999. Their primary customer base ranges from 15–25 in age who have went on to form network-based communities. They have launched MU in China in a partnership with Korea's Webzen in 2003. They are also servicing Blizzard's WOW.

Source: Korea Game Annual Report 2004–2007

available in local Internet cafés. However, these packages are distributed for free, and thus are generally not available in retail stores. This is an important difference from the U.S. market, where online games are usually offered at game retailers, and the client must be purchased before the game can be played.

5) *Cases of Korean expansion into China*

Nexon–Maple Story

When Nexon was developing Maple Story for release in China, it designed the items in the game to cater to Chinese players, offering familiarity that made the game accessible to first-time gamers. Nexon also took note of the fact that Chinese PCs tended to have less memory than Korean computers, and focused on keeping the game's

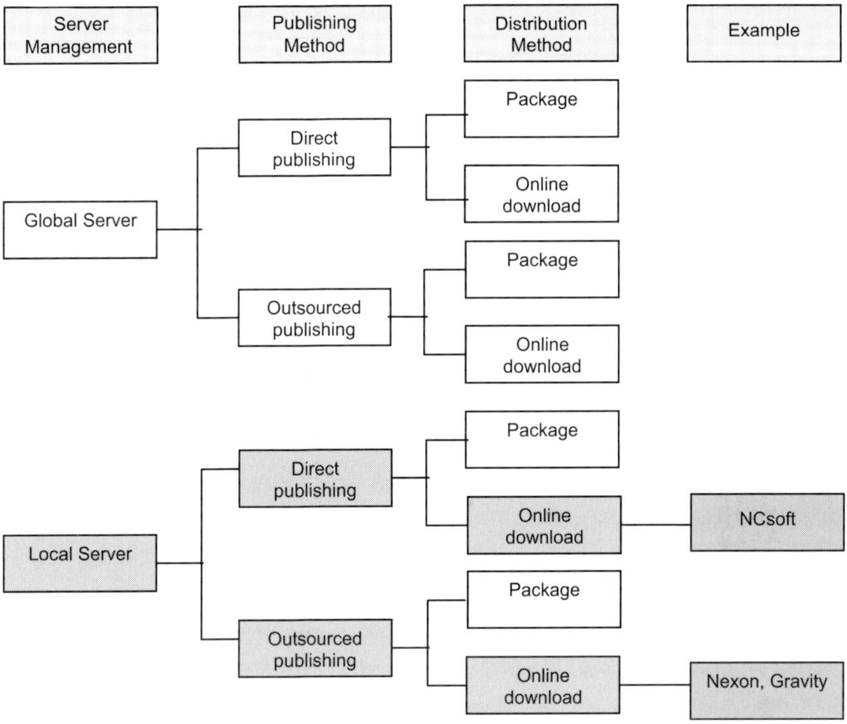

Figure 6.5 Pattern of Chinese market penetration of Korean online games

system requirements low while also making the characters appealing. Nexon launched Maple Story in collaboration with Shanda, and reached 300,000 simultaneous users within two months of release.

Nexon–BNB

BNB has been a casual game sensation, drawing over 700,000 simultaneous users at a time when the market has been dominated by MMORPGs. BNB uses the partial pay model, running an item shop that sells both aesthetic items and items that aid in gameplay. Nexon noted that Chinese players are demographically similar to Korean gamers and that most of them have mobile phones, and has made very successful use of mobile phone marketing campaigns for BNB.

MGame–Yulgang Online

MGame noted that China has a relatively weak communications infrastructure compared to Korea, and when they moved to launch the beta test of Yulgang Online, they did so by distribution 700,000 CDs to local Internet cafés. This circumvented the problem of slow download speeds for their large game client, and provided easy access for both individual gamers and Internet cafés alike. MGame was rewarded for their insight, as Yulgang Online recorded 200,000 simultaneous users six months into the beta test.

Gravity–Ragnarok

The Chinese government has tightened their criteria for rating online games as it has become concerned about the effect that game violence can have on young players. Gravity catered to these concerns, creating character models that looked less violent than those seen in other games. Also, pursuant to requests from Chinese users, Gravity created game maps that included traditional Chinese architecture, which made the game look and feel familiar to Chinese players.

3. The Japanese market and its distinguishing characteristics

By late 2007, roughly 70% of Japanese had broadband Internet access either at home or at work, equivalent to the Korean figures from late 2004. In 2001 the Korean market was experiencing MMORPG fever, led by the amazing popularity of Lineage. Online casual and board games were also experiencing rapid growth, led by Han Game. With broadband penetration in Japan at similar levels, will it experience the same growth in online gaming?

Online games have certainly done well in Japan. Korean games currently dominate the market, led by Lineage 2 and Ragnarok. Ragnarok alone has over 1 million registered users, and has been instrumental in spreading online gaming in Japan.

GungHo Entertainment, the Japanese publisher servicing Ragnarok, went public on the Osaka Securities Exchange in 2005 and reached a market capitalization of $3.3 billion. Compare this to Square Enix's market cap of $3.2 billion and Konami's $3 billion, and it demonstrates the hopeful outlook for online gaming in Japan.

Despite these initial successes, however, online gaming in Japan is unlikely to experience the stratospheric growth that it has in Korea and China. The market is more likely to expand smoothly but gradually, as it has in the United States. The analysis below of the U.S. online gaming market can serve as a reference point.

> "Of the console games geared toward the U.S. market a large number of games are well-made and cater to the ethos of U.S. users. From the perspective of users acclimated to this kind of game, PC-based online games are foreign and demanding. Also, most of them are not trained to play games using PCs. Conclusively, unlike Korea, the American MMORPG market is extremely small, and a majority of the estimated 2 million to 3 million users are hardcore users. Due to such market conditions, EA, an online game service provider of EA, has temporarily withdrawn from the U.S. online game market after posting huge losses until 2003."[3]

The slow pace of online gaming growth in Japan was once blamed on insufficient high-speed Internet infrastructure. But as previously mentioned, broadband access has spread rapidly in the last few years, so slow connection speeds seem unlikely to be the culprit. Furthermore, as Figure 6.6 shows, broadband Internet access is more affordable in Japan than anywhere else in the world (Wi, 2006b).

The sluggish growth of online gaming in Japan must be attributed to other factors. Broadband access is widely and inexpensively available, and yet the Japanese online game industry has still not undergone the explosive growth seen in China and Korea.

One possible factor is the lack of PC expertise and low Internet use among Japan's teenagers. Japan's youth are not sufficiently

[3] Contents Management Institute (2005), p. 135.

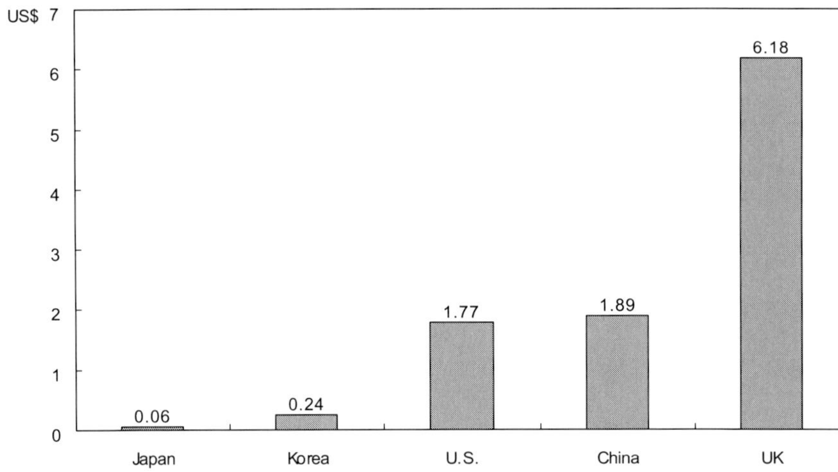

Figure 6.6 International comparison of Broadband usage fees (Unit: USD)
Source: Forecast and status of the Japanese Game Industry (2006 Edition)

trained in computers at school, which affects their PC usage at home as well.

Another likely factor is the widespread popularity of mobile phones in Japan. Japanese teenagers tend to spend hours a day on their mobile phones, talking to friends, sending email and text messages, and playing games. As Figure 6.7 demonstrates, the popularity of mobile phones has had an adverse effect on other entertainment options.

Despite these factors, growth in both PC and Internet use is expanding in Japan, which means a latent market for online gaming exists. Game developers just need to find a way to tap into that market by offering the right kinds of games that appeal to Japanese gamers.

As of 2006, there are 474 online games available in Japan, and 420 million players pay to play them. Korean games have an overall 27% market share, but when console-based games are dropped from the equation and only PC games are considered, Korean online games have a larger share than their Japanese counterparts (refer to the figure below). Korean games in Japan offer both set-amount price plans and item sales.[4]

[4] http://www.onlinegameforum.org/.

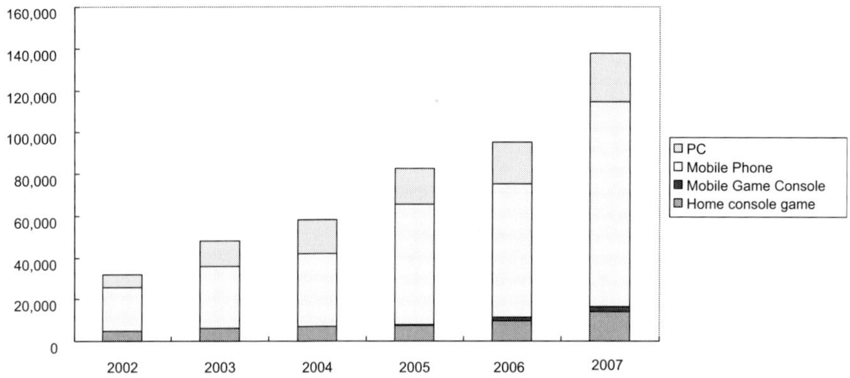

Online Game Platforms	2002	2003	2004	2005	2006	2007
Console Game	4,900	6,000	7,000	7,600	9,500	14,000
Mobile Game Consoles	30	10	40	60	1,700	2,600
Mobile Phones	21,000	30,000	35,000	58,000	64,000	98,000
PC	6,000	12,000	16,000	17,000	20,000	23,000
Total	31,930	48,010	58,040	82,660	95,200	137,600

Figure 6.7 Online game industry by platform (Unit: 1 million Yen)

Source: Forecast and status of the Japanese Game Industry (2007 Edition)

1) *Distinguishing attributes of the Japanese online game market*

The Japanese game market is primarily made up of console-based, offline games. PC and online gamers are currently the minority of Japanese gamers. Japanese gamers who play online games, however, tend to be heavy users. While more than 50% of PC and mobile game users play less than one hour a week, only 26% of online game users do so, and online gamers who play more than two hours a week make up the majority.

The transition of console gaming to online play is another defining attribute of the Japanese console game market. The current market

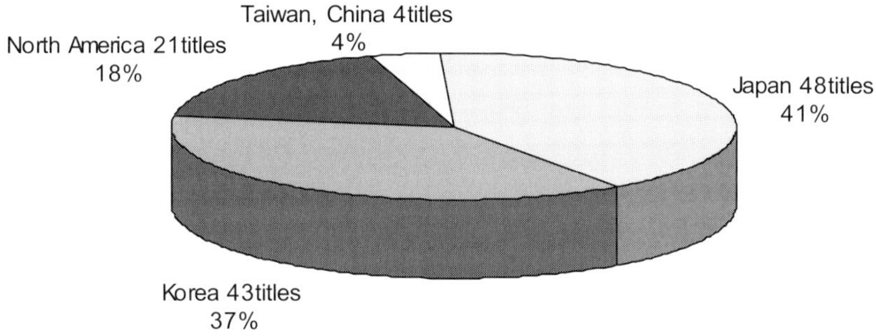

Figure 6.8 Online games serviced in Japan (2006)

Source: <http://www.onlinegameforum.org>

Minutes	1~30	31~60	61~120	121~180	181~240	241~300	301~360	more than 361
PC	32.4%	23.6%	12.4%	9.2%	5.6%	2.8%	2.8%	10%
Online	10.9%	15.2%	8.7%	15.2%		10.9%		32.6%
Mobile Phone(ST)	34.9%	20.2%	17.4%	11%	6.4%	1.8%	1.8%	4.6%
Mobile Phone(N/W)	33.3%	16.7%	11.1%	5.8%	0%	0%	0%	11.1%

Figure 6.9 Gameplay duration of Japanese gamers (per week, in minutes)

Source: 2005 CESA Game Annual Report

for online games is dominated by PC-based games, but as console games begin to offer online play, the market potential is enormous given that console games represent the largest market for console games in Japan. The Playstation 2 offers some network play and the Playstation 3 is to offer far more, which will bring more gamers into the online space, and might also benefit existing PC-based online games.

Also noteworthy is the popularity of online casual and board games in Japan. The target market for MMORPGs is hardcore gamers, and thus it can be difficult for new gamers to feel comfortable playing them. Acclimatizing users to a PC-based game interface via more accessible casual and board games before introducing them to other game genres could prove successful.

Han Game demonstrates the potential of that strategy. With 120,000 simultaneous users and 10 million registered users, Han Game is the top casual game in Japan. These numbers represent a more than twofold increase from 53,000 simultaneous users in early 2004, less than a year before, and show that Japanese gamers will react when provided with attractive content.

2) *Development efforts of Japanese game firms*

Many Japanese game developers are currently working on online games of their own. Online game divisions in these companies face difficulties, however, due to the primary focus on console and arcade games as well as the lack of understanding of online games.

As explained in Chapter 1, both the technology and business models of online and offline games differ markedly. Offline games are created and released, like a movie, and the release signals the end of the game's development. The development team is disbanded upon the game's release and moved to a new project.

With online games, however, the game's release does not signal the end of development but rather the beginning of a new phase. The developer must regularly patch the game with updated content if it is to be successful in the long-term. Many game developers view online games as never-ending, tedious work. Sega's network business director Someya Matsuhiro makes the distinction below.

> "Success in the arcade game development industry is determined by the visuals on the first couple of screens that run for 5–10 seconds. After a quick glance upon entering the arcade, gamers decide whether to insert the $1 coin into the machine. The key is to provide attractive visuals to draw the player in the 5–10 seconds allotted. However, with online games, you have to play at least 3–4 days to find out whether you like it or not, and the success of the game hinges on how to form communities with game friends within the game."

Having to constantly interact with and provide service for users is also seen as a chore by some game developers. In some respects, offline game development is akin to an art form. Developers keep tabs on user trends or shifts in preference prior to development, but once the creation of the game is underway, they can create their masterpiece without worrying about the user.

Online games are far more user-centric. Users are monitored continuously as the developers watch to see what players do and do not enjoy about the game. They then must make content changes to reflect those opinions. Such changes are often contentious and it is generally very difficult to make all of the players of a game happy. For example, if one character class is viewed to be too weak and is thus strengthened, players of opposing classes may object. To many game developers, catering to user demands is viewed as being cowed by their users.

The amalgamated result of these factors is that Japanese game developers have interests that tend to run counter to online games. The Japanese online game industry is a somewhat stagnant niche market, and the online game sector in Japan is not very profitable, even showing losses at times. This coupled with the vast popularity of offline gaming means that Japanese game developers view a switch to making online games as undesirable.

Much of this hesitancy results from a lack of understanding about the potential of online games. This phenomenon is not limited to the gaming industry, but has been spotted in other sectors as well. Clayton Christensen, a professor at Harvard Business School, describes the phenomenon of existing technology and systems repressing new innovation as "the innovator's dilemma." (Christensen, 1997; Christensen and Bower, 1996)

Research into innovation and corporate behavior has shown that corporations with successful track records tend to become complacent and have a difficult time adapting new products, bogged down by existing habits and systems. When rapid innovation takes place, the wide-ranging support for existing systems acts as an obstacle to the adaptation of new technology or products. This is why innovation often comes from small companies. Without an existing product to serve as a barrier, they can concentrate fully on new ways of doing things.

One reason for this phenomenon is that new products are often functionally inferior to old products when they are first released. For example, early notebook PCs were slower than desktop PCs, and their displays weren't even in color. Many computer companies disregarded notebooks at this stage of their development because they expected them never to be able to come close to desktop PCs in performance, and instead focused on enhancing the performance of their desktop machines (Wi, 2006c; 2006d). In other words, companies tend to assess new products using their existing products as the standard, and this can lead them to disregard a new product which might be inferior at the time of its release, but has great long-term potential. This "innovator's dilemma" appears to be taking place within game development companies in Japan.

3) *The strategies of Japanese game developers*

Japanese game developers have pursued many different angles in approaching the online gaming industry. This differs from the development of the markets in China and Southeast Asia, which have evolved in a pattern similar to that of the Korean market. In these markets, PC-based online games and MMORPGs drove the market expansion, followed by casual games that appeal to light users.

The Japanese market, however, is very different from that of Korea, China, or Southeast Asia. Console games are firmly entrenched in Japan, and very popular. Most Japanese gamers are less familiar with PCs than their Korean counterparts, and do not perceive PCs as a gaming platform. Due to these market differences, Japanese developers are employing a wide range of strategies to draw users to their online games.

Bandai-Namco, for example, has developed stores called "Red Zone" in Tokyo, which are similar to Korean Internet cafés. They offer games for online play, as well as games designed for competition between teams of players playing from within the store. Red Zone also allows Bandai-Namco to train users in PC gaming and familiarize

Figure 6.10 Namco's Red Zone

themselves with the genre, to open up new possibilities for games that can be marketed to these users.

Koei is experimenting with transitioning existing console game titles to an online format, which has the advantage of the games already being familiar to players. After releasing Nobunaga's Ambition relatively early, in 1998, they have since introduced Romance of the Three Kingdoms, Aerobiz, Three Kingdoms Battlefield, and Nobunaga's Ambition Online.

In 2005, Koei also looked toward the international market and began developing Nobunaga's Ambition Online overseas, offering the game in China and Taiwan in collaboration with local partners such as Softworld. They also aim to release their title Uncharted Waters in China, Taiwan, Korea, and eventually North America and Europe. Koei has been one of the leading Japanese developers in the online game market; one of the few developers to fully embrace the potential of online gaming.

Sega is another Japanese developer with a long history of online activity. Sega began offering their Game Library online via their Mega Drive console all the way back in 1990, and the Dreamcast, launched in 1998, was the first home console to include a modem for online play. Phantasy Star Online (PSO) was another innovative Sega launch, being one of the first console-based RPGs able to be played online with other players. Sega has since withdrawn from the console market and is solely a software developer now, but their experience with online play has proven beneficial. Currently, Sega is working to import Korean online games to Japan, and also runs SEGAlink, a community website.

In addition to those major firms, game publishers such as GungHo and NHN Japan are following the Korean model and attempting to expand the online gaming market by releasing traditional PC-based online games. The diversity of the different approaches to online gaming by Japanese developers is sure to help grow the market, as there are a wide variety of games either available or in development, and likely something of interest to any gamer. The table below further highlights the plans of major Japanese developers.

4) *Strategies for approaching the Japanese market*

Japanese game developers have been working to assess the market, experimenting with different ideas to see what types of games are well-received by Japanese players. What sort of lessons can be taken from the experience thus far, and what strategies should be considered by developers looking to succeed in the Japanese online game market?

One factor to note is that the Japanese market is very different from the Korean, Chinese, and Southeast Asian markets. Online games are still currently a niche market, with console video games still the dominant player. Also, the popularity of mobile phone-based entertainment acts as competition for the attention of users, and draws away from traditional PC-based online gaming.

Current experiments in Japanese online gaming include Bandai-Namco's Red Zone and Sega's electronic card games Mushikingku and Great Battles of the Three Kingdoms. While Red Zone has taken PC games and moved them to an arcade-style format, Sega has gone the opposite route, releasing their card games in arcades and then adopting them for online play. Sega's card games possess many of the attributes of traditional online games, such as direct competition between

Table 6.6 Strategy of Japanese game companies

Game maker	Market strategy
NHN Japan	They predict that the Japanese market will continue to grow after 2007. In product strategy they are aiming at strengthening existing content and service, with the primary goal of breaking 200,000 simultaneous users. As 2008 is the year in which many firms are expected to expand to Japan, in order to solidify their No. 1 rank, they will focus on their successful existing titles to maintain their position.
Koei	An expanded edition of Nobunaga's Ambition Online was released in December 2004. Koei plans to continue to add new titles as well as offering more events. Beginning with the launch of Uncharted Waters, they will continue to release and service an online action game and Romance of Three Kingdoms Online in 2007.
GungHo	Adopted global publishing as the long-term goal. GungHo is watching all gaming platforms, creating games for mobile phones, PCs, and even consoles. Multiple titles are in development, and GungHo will continue to pursue servicing of foreign titles.
Bandai-Namco	Accelerating spread of online games via a national network of stores to compensate for the lack of location-based game servicing — a drawback of online games — they will continue to publicize the attraction of online games, and game control via mouse and keyboard. Contemplating Counter Strike Neo, a shooting game-based title completely localized to the Japanese user, as their killer title. In the coming months, Namco will aim at expanding the target user groups via assuming a multi-content provider role.
Sega	Poised to promote the latent expansion of the online game market by employing Internet entertainment as a tool with which to expand the user base. To that end, they have launched SEGAlink, a game & community service. A unique service combining various casual games, online games, social networking, a home page, and a blog, users can communicate with others or play games with SEGAlink employing the Internet as a comprehensive entertainment medium. They operated RF Online, a smash hit developed in Korea and also working on developing their own in-house titles.

Figure 6.11 Great battles of the Three Kingdoms game screen

players, user interaction, and creating a sense of anticipation about the next event in the game (the turning over of the next card). The excitement of seeing the next card is similar to the excitement of seeing what a monster dropped after killing it in a traditional MMORPG.

As seen in Figure 6.11, PC-based online games, arcade games, and console games being linked in one networked environment will produce a new style of online game. This concept of multi-platform gaming is distinct from the Korean and Chinese markets, which have focused almost entirely on PC-based games.

Not all genres of online games will be successful in a multi-platform market. MMORPGs, for example, are best suited for being played on PCs with a mouse and keyboard interface, and are thus less likely to be popular in a multi-platform environment. The types of games that have been successful on all three platforms by themselves are the most likely to succeed in a multi-platform market as well, such as first-person shooters and sports titles. The important thing to note is that a multi-platform market means that the genres of games which will be popular will likely change as the popularity of each type of platform shifts.

To further grow the Japanese online gaming market, the user base must be expanded by strengthening the community aspects of online games. Aside from the entertainment provided by the actual gameplay, one large reason for the success of online games is the interaction they promote between their users. Via avatars within the game, players can create alter egos, chat with other players, and make lasting friendships.

This social aspect of online games is an added value for players, and induces them to pay for access to a game. Although free games abound, users have shown themselves willing to pay for quality games with strong social interaction available. Avatar item sales are another example of added value. Players purchase digital items which have no intrinsic real-life value, but add to the entertainment had within the game.

Japanese gamers have shown themselves to be strongly drawn to the community aspect of games. However, in Japan online games and community websites are generally separate entities. Community is an inherent part of online gaming, and if the Japanese market is to expand, they must be drawn together. Online games must have a community appeal to them as well as gameplay. SEGAlink and NHN Japan's Kookooroo service are both taking steps to address this gap, working to combine community with gaming, and are seeing success.

Another element key to the expansion of the Japanese online gaming market is offering accessible titles that new users can quickly feel familiar with and enjoy. MMORPGs are targeted at avid gamers, and generally users new to online games have a hard time grasping the concepts and complicated controls that MMORPGs incorporate. Using casual and Web-based board games to train users and familiarize them with online gaming is an approach that has succeeded in Korea, and will likely succeed in Japan as well. The saturation of Japan's MMORPG market means that expanding the online gaming user base is critical to market success, and casual games can fill that role.

The Japanese market would also benefit from a variety of pricing plans for online games. As is the case everywhere, charging for content on the Internet is controversial, with businesses unsure about the best way to drive revenue from their users. Should the content be provided free of charge and revenue generated some other way, such as advertising? Or should users be charged a fee to access the content, and if this is the case, what services should be provided that constitute a fair and compelling value?

So far in the Japanese online gaming industry, Korean developers have offered just one pricing solution: the set monthly fee. Games go through a free beta testing phase, and when they go commercial, users must pay a regular fee to play the game. This is a high-risk method, though, as upwards of 80% of beta testers cancel their subscriptions when monthly fees are introduced. Oftentimes, going commercial can sink a game completely, when its player base evaporates after the commercial launch. The large number of MMORPG titles available in Japan also restricts the monthly fee option, as there is always another game available to play.

The global trend in the online gaming market is away from set monthly fees and towards item sales to generate revenue. Japan is no exception to this. However, while Japanese users tend to have high spending ability, the market lacks high-quality content to satisfy that demand. The release of games that satisfy the desires of Japanese gamers will likely see great success, as players embrace the game with open arms and open wallets. Another possible revenue source is via in-game advertising, which has already seem some limited success.

Item sale is a very broad category, with everything from purely aesthetic items to functional items that assist in gameplay; and from limited-use items to permanent items that the player can use whenever they want. When designing an item sales model, game developers need to take into account the differences in the Japanese market from the Korean one, and create their items accordingly to appeal to Japanese gamers.

A final key factor in approaching the Japanese market is the need to acquire and keep teen users. Teenagers tend to be the most avid online gamers, as well as the age group most likely to stick with one game for a long period of time. If online gaming is to succeed and grow beyond its current niche status in Japan, the teenage market must be pulled in, drawn away from their current entertainment of mobile phones and console video games. The concept of path reliance, discussed earlier, makes it difficult to draw teens from one type of game to online gaming. But it is a necessary objective, as they represent the largest market group for online games, and without a large base of teen users online gaming will likely remain a small, untapped market for years to come.

5) *The Japanese market and its distinguishing characteristics*

(1) *Primarily MMORPG-based expansion*

The first thing is that the vast majority of Korean games that have been released in Japan to date are MMORPGs. This is a result of the incredible popularity of single-player RPGs in Japan, with a 27.7% share of the game market. This stands in contrast to the U.S. market, where sports and action games dominate the market and RPGs have a relatively small market share. The popularity of RPGs in Japan is similar to the Korean market, and serves as a welcome facilitator for the spread of Korean online games, which are primarily RPGs themselves.

Just because RPGs are popular in Japan, though, does not mean that all Korean MMORPGs will see market success. While in Korea the most popular online games are medieval fantasy titles such as Lineage, martial arts games such as Legend of Mir 2 lead the Chinese market. In Japan, the leading MMORPG is Ragnarok, with its cartoon-animation characters. Medieval fantasy title Final Fantasy XI boasts a user base of 550,000 users, second only to Ragnarok, but as that number includes users all over the world, it is difficult to draw conclusions specific to the Japanese market.

(2) *Introduction of the partial pay revenue model*

Korean game developers are beginning to drop the set monthly fee in Japan and introduce partial pay systems, such as item sales. Currently, about one fourth of Korean games available in Japan

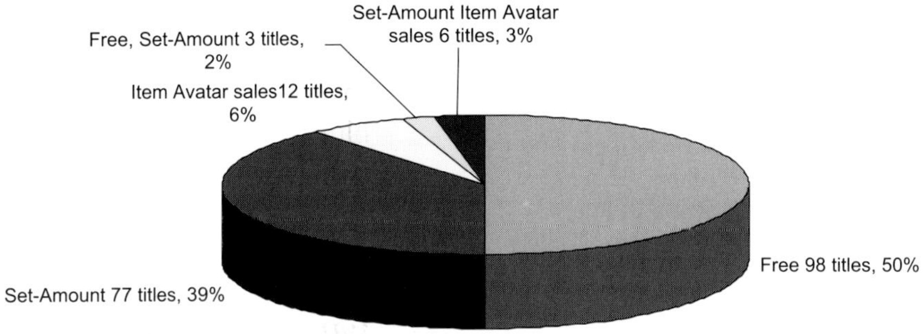

Figure 6.12 Revenue models of Japanese online games

Source: <http://www.onlinegameforum.org/>

operate on a partial pay system. This reduces the resistance to paid games and aims to keep users playing a game for a longer period of time.

One defining attribute of Japanese users is a high latent buying power. With a large disposable income available, Japanese users are willing to pay relatively high prices for a product they enjoy. A survey of an online game forum found that the average monthly fee users paid for access to an online game was roughly $14, but that figure jumped to $43 when item sales were included.

Kim Jong-shin, president of Game On, states that "without placing restrictions, some players purchased as much as $20,000 worth of items, hence we have placed a $300-a-month limit." This information shows that item sales have vast potential in the Japanese market, which is particularly significant in light of the fact that Japanese MMORPG players tend to be reluctant to pay a set monthly fee.

Figure 6.13 is the result of a survey of online game Lineage players in Japan, which asked the amount of money they currently spend to play online games, and the amount they are willing to spend in the future. According to the graph, Korean users rank higher in both areas. Korean users spend $37 per month on average to play online games, whereas Japanese users spend only $23, a full third less. With the national income of Japan being three times that of Korea, it is clear that there is an untapped market opportunity here, and the

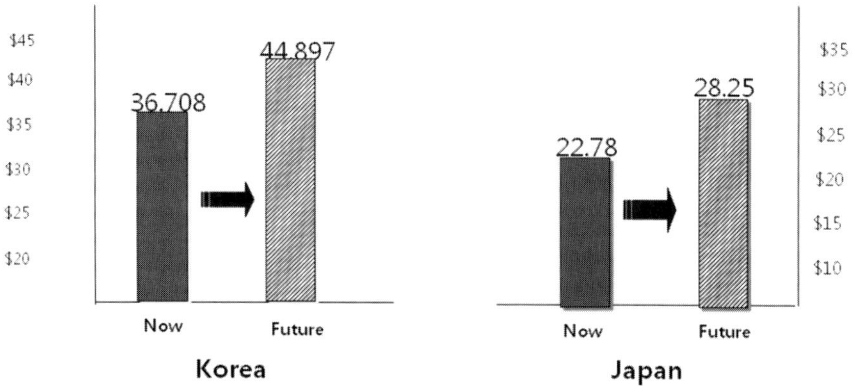

Figure 6.13 Amount paid by online gamer users and future possible payment
Korean users 2,446, Japanese users 2,703

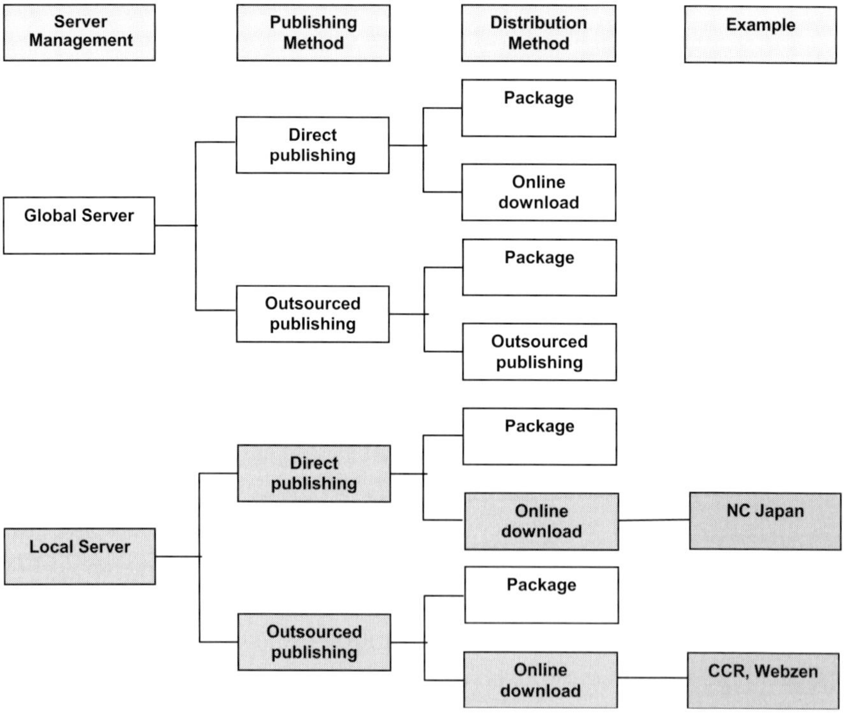

Figure 6.14 Pattern of Japanese market penetration of Korean online games

aforementioned willingness of Japanese players to purchase items is a key way to tap into that opportunity.

(3) *Types of publishing and distribution for Korean online games in Japan*

Korean game developers have primarily opted to distribute their games online, taking advantage of Japan's Internet infrastructure to avoid the cost of retail package distribution.

Most Korean game developers have opted to partner with a local Japanese publisher to release their games. GungHo's servicing of Ragnarok and Sega running RF Online are examples of this type of arrangement. The business deals are essentially the same as in the Chinese market, with a local publisher handling the release of the game and running the servers in exchange for a share of the revenue.

Several Japanese publishers have been working on developing their own titles. For example, in 2004 GungHo created a joint venture with Broccoli, a game developer, to create new online games. Softbank, GungHo's parent company, purchased a 52.4% controlling interest in game developer Gravity via its affiliate company Xer. With this move, Softbank has evolved into a corporation capable of both game development and publishing in Japan as well as being set for foreign expansion.

Some Korean developers have opted to publish their titles in Japan by themselves, without a local partner. NCSoft, Nexon, and Han Game have all taken this route. However, this option is generally limited to the larger developers that can afford the financial outlay to set up a local office, and most small and mid-sized developers have opted to partner with a local publisher to release their games.

4. The U.S. market and its distinguishing characteristics

The U.S. online game market has been exhibiting relatively smooth growth. The overall U.S. game market is led by console games and PC games. The U.S. online game market was $87.2 billion in 2005, but is predicted to reach $264 billion by 2008, passing PC games. The U.S. online game market is widely predicted to maintain steady growth after that.

It should be noted, however, that the numbers for the U.S. online game market encompass all genres of games, including MMORPGs, casual, and board games. Unlike the Korean and Chinese markets, the U.S. MMORPG market remains a small niche. Only approximately 2 to 3 million users play MMORPGs in the U.S., or roughly 1% of the population. Accordingly, growth in the online game market will likely be led by casual and board games that appeal to a broader base of users.

Existing games such as Everquest, Final Fantasy XI, and World of Warcraft have established strong positions in the U.S. market, and new MMORPGs will have a hard time competing against the market leaders. However, broadband Internet access is on the rise, and communications carriers are looking for content to provide on their new high-speed networks. U.S. game developers and publishers have shown a keen interest in Korean online games to fill this role.

The success of Microsoft's Xbox and Xbox 360 sends an important message to the U.S. game industry. Both versions of the Xbox include a hard disk and are capable of online gameplay, and studies have shown that 60% of U.S. console owners are interested in online games. Xbox Live, Microsoft's online gaming service, has seen explosive growth and currently boasts over 7 million registered users, according to Xbox Live division manager JJ Richards. The U.S. online gaming market may still be in its infancy compared to Korea's, but the market potential is clearly there.

1) *Purpose of Internet use*

The results of this study show that PC-based MMORPG players primarily use the Internet for playing games, with 70.64% of respondents reporting as such. PC game users show the same trend in a slightly weaker form, with 54.97% of them using the Internet mainly for games. Users who play online board games show a less dramatic split,

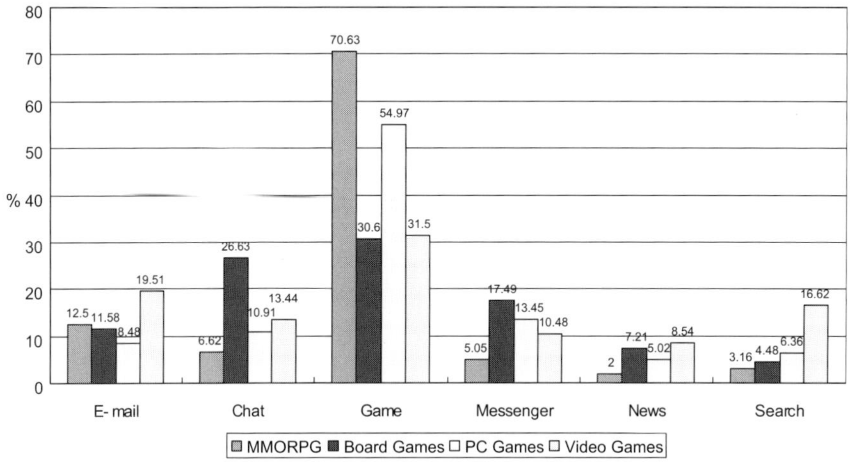

Figure 6.15 Purpose of Internet use of MMORPG players

* $X^2 = 1604.903$, $p < 0.01$

with 30.6% of them reporting games as their primary use of the Internet, but chat follows closely for this group at 28.63%. For all groups, messenger services, email, search, and news come in significantly lower.

Clearly, a large proportion of U.S. gamers enjoy playing games online, but the purpose of going online does differ somewhat according to the type of game the survey respondents primarily play. MMORPG and PC game players overwhelmingly use the Internet for games, clearly demonstrating the market potential of online gaming in the U.S.

2) *U.S. market penetration of Korean game developers*

Korean game developers were offering more than 50 games in the U.S., with most of them being MMORPGs. Major players such as Gravity and NCsoft have established local offices and are directly publishing their games, but mid-sized developers are expanding overseas via GameNGame.com, a game publisher operated by the Ministry of Information and Communication's Korea IT Industry Promotion

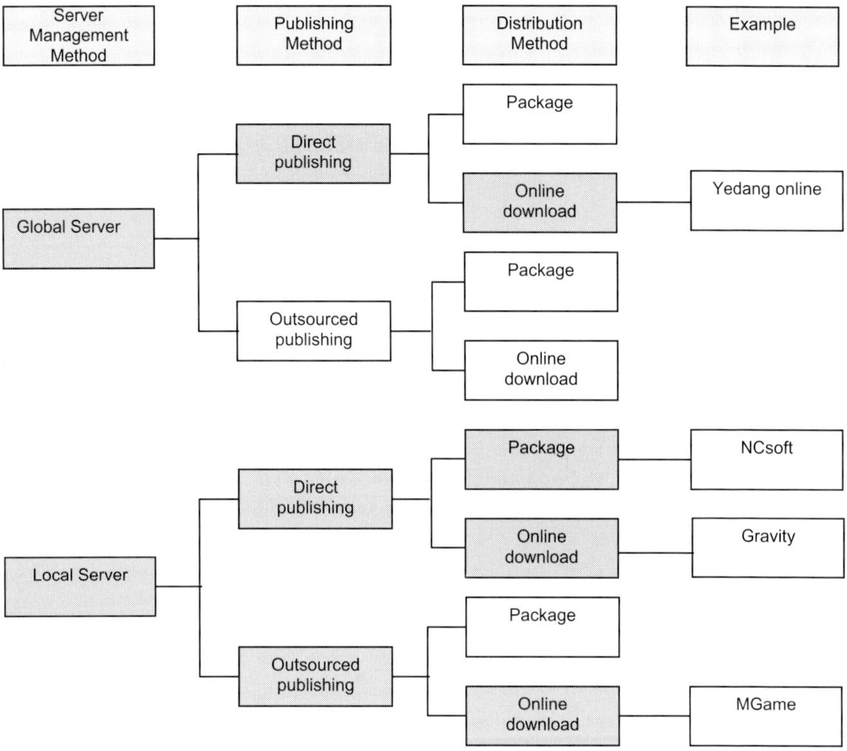

Figure 6.16 Pattern of U.S. market penetration of Korean online games

Agency (KIPA). KIPA built GameNGame.com to act as a global test bed providing skills, marketing, and infrastructure for overseas testing of domestic online games.

Unlike in the Chinese and Japanese markets, Korean game developers offer a wide variety of distribution methods for their games in the U.S. They have primarily focused on four different distribution options, which will be examined in more detail. The methods are:

1. Global server — direct publishing — online download
2. Local server — direct publishing — package sales
3. Local server — direct publishing — online download
4. Local server — outsourced publishing — online download

(1) *Yedang online (Global server — direct publishing — online download)*

Yedang online, founded in 1999, launched its first title "Prestone Tale" both domestically and internationally in April 2000. With 250 employees, Yedang online offers its game in seven countries: the U.S., China, Japan, Taiwan, Thailand, the Philippines, and Vietnam. Prestone Tale boasts 3 million registered users, and is Yedang online is reported as being a profitable company.

Prestone Tale is run on a global server, located in Korea and accessed via a proprietary UDN (Universal Design Network). Players use the UDN to connect to Yedang online's Korean server directly, wherever they are in the world. Kim Nam-cheol, a business development executive at Yedang online, offers this explanation for establishing a global server as opposed to using local servers.

> "Yedang online created a global server because we realized we need to build a global server as in the previous profit distribution model, the license model faced fierce competition. For example, if you operate a global server you can bring a pool of 10,000 to 20,000 users from principal countries, then as an independent incorporation, conversion and management becomes a lot easier. In other words, we're aiming at first securing a user database."
>
> (Interview with Nam-Cheol, Kim, CEO)

Currently, 26% of Prestone Tale's players come from North America, with 22% from the U.S. and 4% from Canada. Most of the players are youths, and the eastern regions of the continent show the highest rates of gameplay.

Separate development teams for global and local servers have been formed, with content updates being performed by different groups of development personnel. This means that patches are often applied at different times to different servers. For example, if four patches are performed on the local server, the English global server might be updated with just one patch at that time with the other three set to follow at a later time.

Yedang online's patches are performed via a dynamic peer-to-peer (P2P) system. For example, if one player downloads a new patch directly from Yedang online's server, another player can then download the patch directly from the first player via the P2P system. On patch days, Yedang online first distributes the new patch to forums and friendly community sites to prevent an overload of its own network. As users download the patch, the available sources for other users to download from grow larger. This allows patches to be distributed quickly across the globe without the need for localized download points, and also limits the bandwidth draw on Yedang online's network.

Billing for Preston Tale is primarily via credit card or prepaid game card. Direct deposit is available via a website called Pay By Cash, but as direct deposit is a costly service, only annual payments can be made in this way. Roughly half of all payments are made via credit card, and most of those paying with credit cards are adults. Due to problems posed by identity theft, credit card authorization is granted only after a rigorous authentication process.

Prepaid game card sales are becoming more popular in the U.S., and Yedang online has recognized this opportunity. Prepaid game cards are sold offline via local retailers such as game stores and convenience stores, and allow users without credit cards to play the game. Kim Nam-cheol of Yedang online states the following regarding the potential of prepaid cards.

> "For example, paying in package deals in conjunction with ADSL fees requires a lot of time to change user conception but when users of a latent market encounter prepaid cards at offline they accept it with familiarity. That is why the prepaid sales market is bound to grow in the future."
>
> (Interview with Nam-Cheol, Kim, CEO)

Marketing in the U.S. has proven to be Yedang online's greatest hurdle. The company does not have the resources to fund a nationwide marketing campaign, as the U.S. is simply too large and Yedang online is too small. Currently, they are engaging in targeted marketing campaigns in specific cities rather than the U.S. market as a whole, and the strategy seems to be paying off as their subscriber numbers continue to rise.

(2) *NCsoft (Local server — direct publishing — package sales)*

Founded in 1997 and boasting hugely successful titles such as Lineage, Lineage 2, and Guild wars; NCsoft has established itself as the leading online game firm. With 2,500 employees, NCsoft primarily services MMORPGs. When it expanded into the U.S. market in 2000, NCsoft recruited star developers such as Richard Garriott of Ultima fame for local game development, to create City of Heroes and Guild Wars. NCsoft also purchased ArenaNet, a Washington-based game developer, to help in its U.S. expansion efforts.

NCsoft's U.S. releases include Lineage, Lineage 2, City of Heroes, Guild Wars, and Auto Assault. To date, City of Heroes has sold over 180,000 packaged copies, and Lineage 2 has sold over 96,000 copies in the U.S.

NCsoft is truly a global company, offering games in the U.S., Taiwan, Japan, China, Thailand, and Europe. They have local offices in the U.S., Japan, China, Thailand, and Europe. Their estimated 2005 revenue is $300 million.

NCsoft's pricing plan for the U.S. market is different from those it offers in Asia. In the U.S., NCsoft's games are sold in packages at retail outlets, and include passes for one month of free gameplay. After the first month, players can either subscribe via credit card or purchase prepaid game cards to continue playing.

NCsoft's games are sold at major U.S. retailers including Walmart and Gamestop, which include traditional retail marketing such as elaborate store displays and on-site events. Games are also sold online at PlayNC (http://www.plaync.com) and via online retailers such as Amazon (http://www.amazon.com). NCsoft does not offer online downloads of their games in the U.S. market.

NCsoft's development process in the U.S. goes beta test → commercialization → localization. Lineage went commercial after six months of beta testing, and as it was produced in multiple languages from the outset, no separate localization needed to be performed. Patches are also able to be applied after undergoing a simple translation process. And unlike most other Korean online games released in the U.S., NCsoft provides a manual to help new players get acclimated quickly.

In terms of server management and operations, while NCsoft is wary of hacking in Asia, it does not view hacking as a significant threat in North America. Kim Ju-young, director of promotions at NCsoft, has the following to say regarding service management.

> "In the U.S. online game service has become systemized to a degree hence no serious problems are expected to arise. U.S. may be ahead in terms of game engine development but in terms of reflecting distinct player traits in the game, and due to the fact that we have experience servicing games, NCsoft is on top."

NCsoft operates local servers and has a full satellite office in the U.S., which is significantly more expensive than partnering with a local publisher. Problems can also arise with delayed transmission of new policies and patches between headquarters and the satellite office, but the advantage is that they retain full control of their game and also get to keep all of the profits.

NCsoft's U.S. market strategy follows the local server-direct publishing-package sales model, which is an expensive option, but more lucrative than other distribution methods if successful. NCsoft's size and financial resources allow it to take this route, enabling it to establish local offices in any market it wishes to enter as well as recruit local talent and buy out domestic developers to assist in their releases.

(3) *Gravity (Local server — direct publishing — online download)*

Founded in 2000, Gravity is a game developer poised to penetrate the global online game market with its smash hit Ragnarok. With players in 25 countries including the U.S., Japan, and Taiwan, Gravity primarily offers MMORPGs but also dabbles in mobile phone, shooting, action, and puzzle games.

Gravity's U.S. expansion followed the local server-direct publishing-online download path of distribution. Gravity established a satellite office in Los Angeles in March 2003, and an American staff overseas the local servicing of Ragnarok. Their Game Master team is made up of all Americans who are provided with 6 months of training, which allows for reliable customer service, free from concern about language barriers.

As in Korea, Gravity's games are only offered via online download. The offline retail package market is still large, but Gravity believes that the market is transitioning to online downloads, and the pace will only accelerate in the future. Kang-Han-keun, an executive director at Gravity, states the following.

> "Employing packages as promotion material is highly desirable. No matter what an aggressive ad campaign we execute, the actual players are not exposed to them but are exposed instead to advertisement we put up in the game departments of WalMart or game retailers. Because we employ downloads, that part is off-limits to us. We get a return when we expend sufficient marketing/promotion costs, but if we sell packages and are passive in promoting it, the promotion costs will just fall through. Of course we can execute promotion campaigns in conjunction with the package release but that only applies to special editions. If we estimate 100,000 users, it's a way to produce just 20 to 30 thousand copies for sale. It's not making a starting kit, but including various items aside from the CD in the package, and selling it has a limited special edition."

Ragnarok's bright graphics and simple interface can be quickly picked up by anyone, and this accessibility has brought it great success. Gravity's target users are casual gamers on the verge of crossing over to MMORPGs, who find Ragnarok engaging and yet easy to understand. A game's community aspect is important to U.S. users, but less so than to Korean gamers, as U.S. users are less likely to hold offline meetings. Playing with a social group is still important to U.S. players, though, and 60% to 70% of Ragnarok's users belong to a guild in the game.

Gravity has faced challenges due to the differences between the U.S. and Korean markets, and Korean developers looking to expand to the U.S. should take heed. One example is in the way items are distributed in the game as a result of combat with monsters. The rules for how items are doled out are different between the U.S. and Korea.

> "A boss monster exists in our game and when characters capture it, at first we offered rare items to the character who landed the most blows on the monster. The same rule was adhered to in the U.S. and we had no problems. Users in Korea complained why not award rare items to the character who lands the first blow so we changed the rule to reflect their suggestion, then users in the U.S. objected but it was too late to revert to the original setting."

A final attribute of Gravity's U.S. expansion strategy is their marketing strategy, focusing on the unique traits of Ragnarok compared to other MMORPGs and including magazine ads featuring Korean Internet cafés, highlighted as pioneering the MMORPG market. Gravity is also focusing its marketing efforts on existing online gamers rather than mass marketing to people who have never played an online game.

(4) *MGame (Local server — outsourced publishing — online download)*

MGame was officially incorporated in 1999, but has been developing online games since 1996. Their offerings span genres from MMORPGs to casual games to board games, including titles such as Neo Dark Saver, Myth of Soma, Droiyan Online, Online Worms, and Online Bomberman. MGame established a satellite office in Japan, but in other foreign markets they have chosen to partner with local publishers.

With annual revenues of $300 million, MGame has expanded to 10 countries including China, Taiwan, and Japan. Domestically, MGame offers a wide variety of games via the MGame portal.

MGame's U.S. expansion has followed the local server-online download-outsourced publishing method. MGame predicts that broadband service will continue to grow rapidly in the U.S. market, and thus they prefer to offer their games for online download rather than retail package sales. MGame is a well-known Korean developer but their brand name is relatively unknown in the U.S. market, thus they have chosen a collaborative strategy with a local publisher to release their games. MGame is also attempting to collaborate with high-profile firms such as AOL and Yahoo! to market their titles.

To minimize any reluctance on the part of U.S. users to pay for their games, MGame has outsourced its billing to a credible U.S. firm. The issue of paid gameplay is an important one in the U.S., with some users hesitant to give their credit card information to a foreign company. Overcoming this problem is key to success in the U.S. market. MGame Director of Foreign Business Development Choi Seung-hoon has the following to say.

> "We have faith in the possibility of ISPs rather than packages. That is why we forged a partnership with K2 Network. They will secure a Knight Online user base with service via ISPs and provide proof that ISPs can assume a central role in the industry. It's safe to say we are undertaking this current task as a landmark for our future expansion."

In addition to one primary U.S. publisher, MGame is looking into offering their titles via individual ISPs as well, which offers the advantage of allowing users to simply add on the fee for gameplay to their ISP bill. This further helps to limit any hesitation on the part of U.S. users about giving MGame their personal information. This method has already proven feasible in the U.S., with AOL's servicing of Matrix Online, in which users can have the monthly fee put on to their AOL bill. MGame believes that this model will become the norm, and has been working to partner with ISPs to offer its titles. However, ISPs tend to be concentrated on the East coast and in Silicon Valley, not quite a wide enough spread to cover the entire U.S. Choi Seung-hoon says.

> "Since finalizing the licensing contract K2 Network has managed parts of marketing and service. While K2 Network handles server

operations, GM functions and customer service, MGame dispatches related personnel for joint management."

If a game developer can partner with a portal such as Yahoo! or AOL, many marketing opportunities are opened up, such as providing game reviews and information on the portal site. The portals can also be used as an intermediary for gameplay and billing, and discussed previously. There are currently about 200 ISPs in the U.S., with smaller ISPs in every state ranging from 10,000 to 20,000 subscribers. MGame plans to launch Knight Online with these ISPs to expand its market reach.

MGame's assessment of the U.S. market comprises two broad elements. One is the distinct attributes of U.S. games, and the other is the distinct attributes of the users who play those games.

MGame has concluded that U.S. gamers are partial to games enabling cooperative or competitive play such as Starcraft or Diablo over MMORPGs. U.S. gamers tend to prefer playing with a smaller group of players as opposed to the thousands of players on an MMORPG server. Further, MGame believes that concerns about hacking lead U.S. players to prefer games with a small group of trusted friends as opposed to thousands of strangers.

MGame's Director of Foreign Business Development, Choi Seung-hoon, compares the content of Korean and U.S. online games below.

"While in Korea users plunge into the game early with the game boom, in the U.S. avid gamers make up the primary user base, and they are logical users. In Korea many quests are short-term, but in the U.S. there are long-term quests, some of which take one year to complete. EverQuest awards valuable items once a year, and this is a form of user management. In order to obtain the valuable items, users become immersed in the game for years on end. In terms of game quality Korea does well in the beginning but with the passing of time, they grow wanting in understanding user mentality. They continue to create short-term quests in large volume due to lack of production ability, mathematics skills, function and logic skills, and

> fail to produce year-long quests as offered in EverQuest. When creating quests language, language differences and historical basis should be thoroughly examined. In this regard U.S. game makers take a more painstaking approach in quest production."

In terms of game content, Mr. Choi also noted that U.S. users tend to be very logical and analytical, and unlike their Korean counterparts, interested in long-term quests. Thus far, Korean games have not adapted to these traits, but game developers looking to release a game in the U.S. would do well to take note of them.

Chapter 7

International Comparison of Online Game Users

The online game is not a simple product but a new form of human society. Online games evolve with the player, in line with the respective cultures and societies. Accordingly, the same game differs by country and culture, and even in a single region the culture and rules are modified according to servers.

For example, in case of Ragnarok, an open market exists wherein players can set up stalls. Player behavioral pattern for setting up vending stalls differs by country. With Korean players if the first user erects a stall at the market center or other prime locations, subsequent users choose among the locations available. In other words, the rule adhered to is an unmethodical first-come-first-served when it comes to vending stalls arrangement.

However, Japanese Ragnarok users display a different behavioral pattern. Users who enter the market first set up stalls in the remotest corners and others follow suit by setting up adjacent stalls circling the market. If the circle is complete, a smaller circle of vending stalls is created within the larger one. In other words, vending stalls are created in concentric circles.

The siege in Japanese Lineage is also different. In Korea the siege is referred to as the highlight of Lineage. A siege involves occupying castles and the guild that does also acquires honor and dignity. That is why guilds within Lineage prepare for sieges ranging from a few days to months. A preference for sieges also applies to Chinese users.

However, Japanese users do not prefer sieges. The following anecdote is a case in point. Korean users were engaging in gameplay on a Japanese Lineage server. As they did in Korea, they formed a guild, and eventually occupied all five castles in the game. Such predominance was cause for complaint from Japanese users and a coalition of Japanese guilds waged a siege. Most Japanese users playing on the server took part in the coalition hence they triumphed without much effort. After having occupied the castles, Japanese users returned each castle to the game maker, and the castle-occupation competition was initialized. Of course, Japanese users did not show much interest in castle occupation following that incident.

The anecdote above indicates cultural disparities among online game users, and the behavioral pattern that stems from such cultural disparities. Then let us examine the differences and similarities among users in terms of region and country, and produce a set of distinguishing attributes for each group.

1. Community identity

In Chapter 1, I noted a fundamental defining attribute of online games is community identity. Community identity is a characteristic exclusive to online games, which produces sentiments of racial solidarity, camaraderie and fraternity via intra-game communication between users. Online game players cannot engage in gameplay alone, and in the game the player carries on another form of life in a virtual society.

However, users in the U.S. or Japan, where console game markets have been firmly established, are not familiar with intra-game community identity. From childhood, they are acclimated to playing video games alone. As such, users acclimated to video games cannot become online game innovators or early adaptors. Accordingly, in markets of developed countries groups which prefer video games and online games exist separately.

According to the results of my study on Japanese Lineage users, 79.7% of them limit their gameplay to online games.

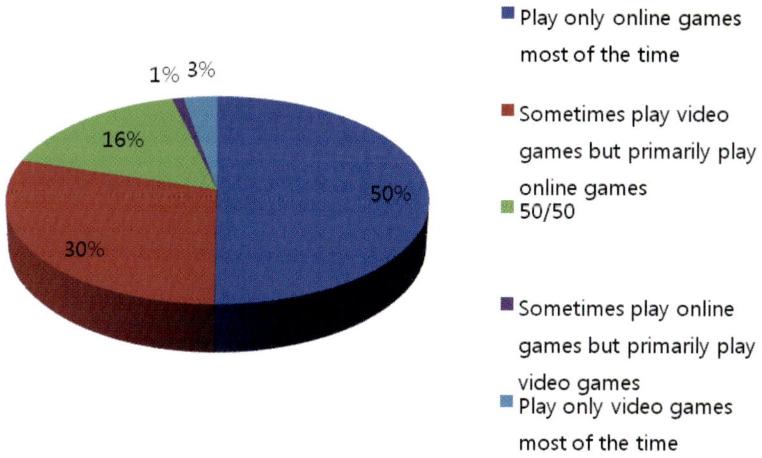

Figure 7.1 Simultaneous rate of use of console and online games

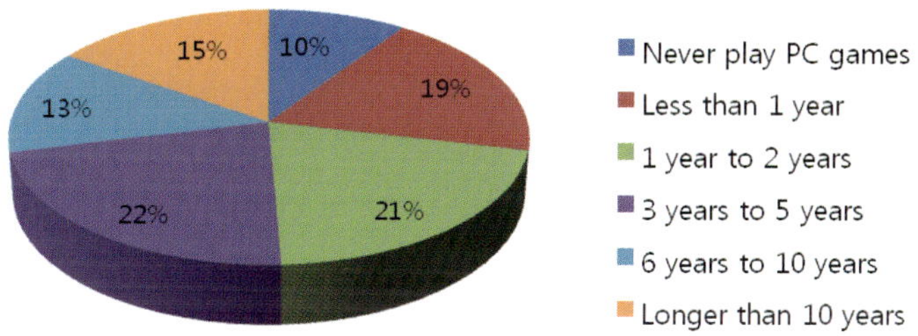

Figure 7.2 PC game use frequency rate of Lineage users

When asked the proportion of video games they play, more than half of the respondents responded they only play online games.

The users who engage exclusively in online gameplay are evolved former PC game players. According to the Table 7.3 below, more than half of current online game users have PC-gameplay experience of longer than three years. In contrast, Japanese video game users have experience playing Nintendo's Pamicom or Sony's PS and/or PS2, but they have not played PC games before. As such, video game users and PC game users adhere to different evolutionary paths.

The results in Figure 7.2 indicate the division of online and video game users in markets of developed countries such as U.S. or Japan.

Perhaps online game users of developed countries can be deemed as a group possessing an intense drive toward the exclusive online game trait of community identity. In other words, early adaptors of U.S. or Japan, who have gained exposure to online games first, display a high preference for community identity. Let us examine this issue further by comparing Korean and Japanese online game users.

In Korea online games spread before video games could establish a market.[1] Korean online game users had no choice to make a judgment call vis-à-vis community identity, and were forced to accept online games. Consequently, it can be predicted that among Korean online game users, separate groups which prefer video and online games co-exist. Japanese online game users, comprising users who prefer purely online games, possess a higher preference for community identity than their Korean counterparts. Let us compare the difference in community identity between Korean and Japanese users in Figure 7.3.

Figure 7.3 reflects a difference between Korean and Japanese users in the community identity category. Japanese users scored higher in all categories than their Korean counterparts.

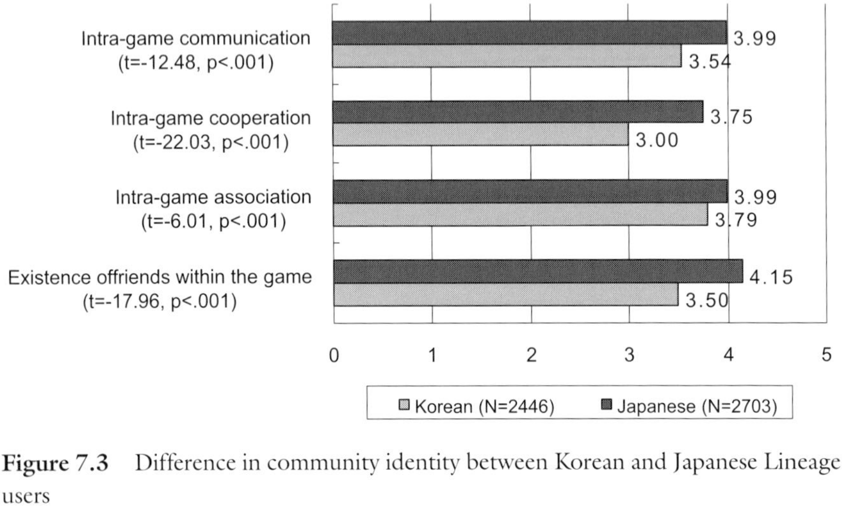

Figure 7.3 Difference in community identity between Korean and Japanese Lineage users

[1] Korea officially began importing game consoles in the fall of 2002.

Continuing gameplay due to game friends and association between friends in the game

The existence of friends within the game has a significant influence on gameplay duration. While amusement derived from gameplay *per se* takes precedence with video games, in case of online games, the existence of game friends substantially contributes to elevating satisfaction derived from gameplay. In other words, if a community is formed, users switching to other games can be prevented, and extended gameplay can be generated. Accordingly, online game users associate with intra-game friends more often than their video game counterparts, and the degree of influence game friends have on extending gameplay duration is larger with Japanese users than Korean users.

These trends also apply uniformly to users in U.S. and China. In case of Chinese users, they have weaker community identity than Korean users. Consequently, sometimes the community is used subordinately in game play. When asked the purpose of community enrollment in the game, 71.13% responded "to facilitate gameplay," 11.09% responded "to talk with fellow community members," 9.4% responded "for protection from PKs," 4.66% responded "dignity derived from belonging to a prestigious guild," and 3.72% responded "association outside the game."

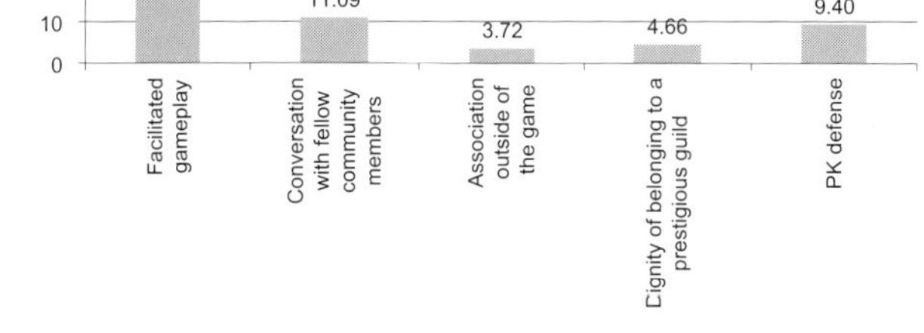

Figure 7.4 Purpose of joining guilds (China)

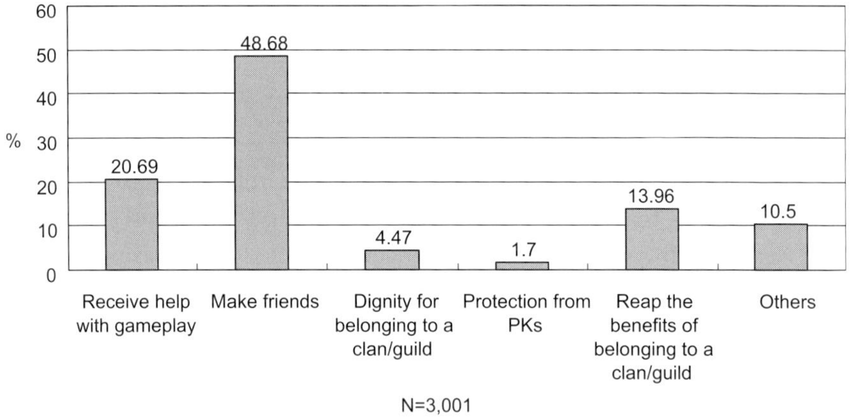

Figure 7.5 Purpose of joining guilds (U.S.)

With these results it can be inferred the primary purpose of Chinese online game user guild enrollment is to facilitate character level ups.

If we look at results from the study conducted on U.S. users "make friends" scored highest in the purpose of joining guilds, with 20.69% responding "receive help with gameplay," 13.96% responding "reap benefits of clan/guild membership, and 1.7% responding "protection from PKs."

The number of respondents who answered they were joining guilds to make friends far exceeded that who answered they were joining guilds to receive help with gameplay. This suggests U.S. MMORPG users possess an intense desire to make friends in the game via guilds as communities.

To sum up, as regards purpose of joining guilds, while level ups employing intra-game communities are a priority to Chinese users, to U.S. users intra-game communication *per se* took precedence over level ups. In other words, while the guild per se is the objective for U.S. users, they are a means to an end for Chinese users.

Intra-game cooperation

In online games the character exp climbs via activities such as monster hunts. The user can set out to increase his experience point alone or by a group effort. Here the user who values communities engages in gameplay in cooperation with other users.

Raids are such joint hunts or battles. This involves tens to hundreds of users who level a joint attack on monsters or wage battle with an opponent country. Below is an interview with a Korean guild, the first in the world to capture the most powerful monster, Onyxia, in WOW (World of Warcraft).

<Interview>

Q: When did you capture Onyxia?

A: After three attempts beginning on Friday at 7, we captured it at 10. We captured it after three total challenges and we will never forget the moment of capture.

Q: Incredible. Do you all belong to one guild?

A: The Chosen Guild has a long tradition; it existed since EQ. As regards unity or affection for the guild, the guild members come second to none. Of course we belong to one guild.

Q: What did it feel like to capture Onyxia?

A: We were brimming with resolve when we launched our final attack. Since we waged scores of attacks after devising strategies, we had unwavering faith that we would triumph. And when it began to decrease to 25%, then to 1/4, guild members reacted by saying it's unbelievable and encouraged each other. Attack squad members and guild members became one in executing the raid and Onyxia finally fell.

Q: Is there advice you would like to proffer other guilds planning Onyxia raids?

A: I'm sure myriad guild members and users are exerting an effort at this very minute to take down Onyxia. Do not fear failure, accumulate experience and analyze from experience, and equipped with agile maneuvers, one day you will triumph.

Source: <http://www.inven.co.kr>

Users participating in a raid must deliberate strategy prior to execution, and receive a prior description of one's role in the raid.

And in actual battle, freedom of individual action is limited to an extreme, and each user must be faithful to his assigned role. Despite these conditions, the user who has a high preference for community identity opts for cooperation with other users in the game.

Users' bent for cooperation in the game was higher in Japanese users, who showed a higher preference for community identity, than their Korean counterparts.

Intra-game communication

In online games the chat window allows real-time mutual communication between users. Through chats users can offer game friends praise or encouragement. Mutual communication is also critical in inter-guild battles. For example, when the energy (blood) level of warrior characters, who are assigned battle duties, drop they must replenish it quickly with the help of mages. If not, warrior characters will die and drop exp and valuable items in possession. Accordingly, when faced with such conditions warriors must transmit help signals to other players without delay.

A large number of users engage in everyday conversation instead of battles in online games. 30% of Lineage 2 users engage in conversation with other users and have no interest in level ups. An example of such a guild is the "Married Users Clan." Gameplay for these married users consists of engaging in conversation on common topics at scenic spots. Topics of conversation usually range from family and children to relationship issues.

Accordingly, the disposition for communication is higher in Japanese users than their Korean counterparts.

As seen above, Japanese users scored higher in all components of community identity ranging from continued gameplay induced by the existence of game friends, intra-game cooperation and intra-game communication.[2] A manifestation of such disparities in community

[2] Wi and Nojima (2003) have classified online game user groups into two camps: one that enjoys association with friends more (values community identity), and one whose priority is enjoyment derived from the game *per se* (values gameness). According to their studies, users who value community identity exhibit a longer gameplay duration than those who value gameness, plays one game longer, and is more willing to pay in connection to gameplay.

identity between Korean and Japanese users is online games spreading before a video game market could take shape in Korea, resulting in players who prefer video games to be absorbed by online games, and who now co-exist with players who prefer online games. In the Japanese online game market users who prefer video games were separated from the online game market resulting in more vivid attributes displayed by online game users.

One distinct fact here is Japanese video gamers not switching to the newly added online games. This indicates the video game market, which appeared to comprise a homogeneous group of users at cursory glance, is truly composed of users with disparate preferences. This fact provides an important perspective in devising strategy for the spread of online games.

In case of the Japanese market users trained in gameness and those who prefer community identity co-exist, and they are divided into different markets. As examined in Chapter 5, it is difficult to naturalize users trained via path reliance to new markets or products.

Then what is the strategy that could absorb the Japanese users swayed by path reliance into a new market? Providing online game training opportunities is the method by which Japanese video game users can be more aggressively steered to the online game market. Japanese users, who are satisfied with exceptional gameness of video games, will not voluntarily switch to online games.

Accordingly, via diverse channels, corporations must afford them experience with mutual communication (community identity) between users, an attribute exclusive to online games. Through such experience it is important to heighten the expectation of online games as a new game form.

Of course, it is also important to aggressively encourage inter-user communication, guild creation and activity as community identity is a higher priority to Japanese online game users than their Korean counterparts. To that end, a communications system far more diverse than that existing in online games being serviced in Korea should be constructed.

2. Game information acquisition channels

There are four common sources of information for users: market-controlled, personal, empirical and neutral.

Advertising, salesmen, packaging and in-store information comprise market-controlled sources, and personal sources include family, friends and colleagues. When consumers first gain exposure to or harbor interest in a specific product, they tend to employ market-controlled sources in the early stages of decision-making and personal sources in the later stages. Depending on the case, consumers obtain information by a trial use, which can be deemed discovery of information from empirical sources. Published materials by government agencies, the media or broadcast news comprise neutral sources, but consumers tend to assign more trust to sources other than market-controlled sources.

Of these sources of information, advertising is widely held to be the most important, especially in case of consumer goods. According to research conducted on information sources, 48% of small-appliance buyers, 35% apparel buyers, and 45% of food buyers responded advertising as the primary source of information (Robert, 1971).

However, online games are vastly different from consumer goods. Personal resources are the most important in online games, with recommendation from friends and online friends wielding decisive sway

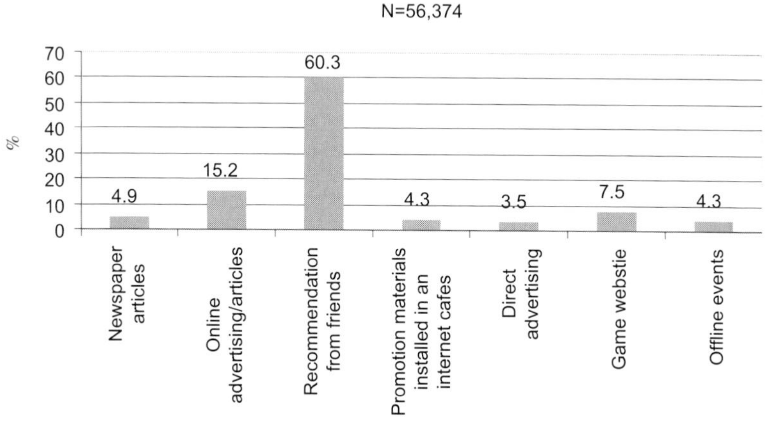

Figure 7.6 Initial method of exposure to game (China)

in game selection. And this influence applies across-the-board to all online gamers worldwide. Let us examine Figure 7.6. This figure reflects results of a study conducted on Chinese gamers.

When asked the method of initial game exposure, 60.3%, the highest proportion, responded "recommendation from friends," 15.2% responded "online advertising and articles," and 7.5% responded "game Web site." The results indicate recommendation from friends is the primary determining factor of game selection. To the contrary, market-controlled sources such as online advertising or articles, newspaper articles or flyers are relatively losing clout as factors swaying game selection. In other words, while personal sources are important channels of information for online game users, they are different from existing channels. It signals the emergence of a new category, the online friend, different from family, friends and colleagues. Depending on the case clout wielded by game friends online is much more influential than that of ordinary human relationships.

This phenomenon uniformly manifests in markets dominated by video games such as the U.S. market. Figure 7.7 reflects the results of a study on information channels conducted on U.S. online game users.

In this study, influence by friends was divided into friends and online friends. When information channels of MMORPG players in the U.S. are examined, 25.9%, the highest proportion, responded "online friends' recommendation," 24.6% responded "friends'

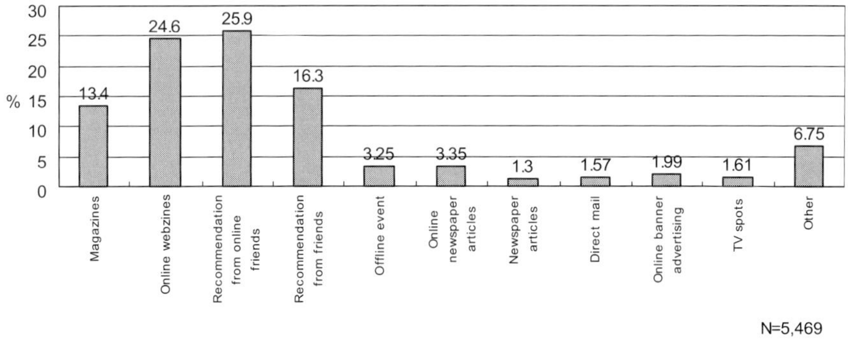

Figure 7.7 MMORPG information channels (U.S.)

recommendation," 16.3% responded "online webzines," and 13.4% responded "magazines." In summary, 51.5% of all U.S. respondents obtained information from friends equal to the results in the study conducted on Chinese users.

However, a notable attribute is the influence wielded by online webzines/magazines. I have already pointed out that in markets of developed countries dominated by the video games, online games have not transcended the niche-market stage, and that video game and online game users exist in separate camps. Nevertheless, online webzines/magazines are a factor in information gathering of U.S. online game users due to the influence of existing PC and video games. Existing PC games were routinely introduced and strategies recommended via online webzines/magazines. Kang Han-keun, a managing executive at Gravity, and Kurosawa, a producer at Square Enix U.S.A., comments as follows:

> We use online forms of advertising to boost our corporate name identification. Then we launched a promotion campaign in magazines targeting animation users. (Kang Han-keun)

> We advertise both offline and online but we primarily employ game magazines. This serves as a method of exposure to users unfamiliar with MMORPGs or untapped users. (Kurosawa)

Accordingly, it can be inferred PC game users reacted to online game advertising on these online webzines/magazines and made the switch to MMORPGs. Such magazines are sold at booksellers and newsstands, and with a high subscription rate, game makers are employing them as marketing tools.

Also, offline events scoring relatively low is a reflection of regional attributes specific to the U.S. market. Gravity's Kang Han-keun comments on this phenomenon below.

> U.S. users are fond of communities, but unlike their Korean counterparts, they are reluctant to meet in person through those communities. They prize anonymity. Even if were to provide an opportunity, we believe almost no one will show up. They are wary of offline meetings, and naturally believe that none will be held. As the region

> is so expansive, they cannot dream of actually meeting. We once held a Ragnarok U.S. championship match, and we provided airfare to Korea for 18 individuals. They informed us that it was their first time to participate in an offline event. They had played together for at least one year, yet none of them had met each other before. In Korea they would call or send pictures, but U.S. users do not call and only associate within the scope of the game; outside of the game, they are stark strangers. That is what is different from Korea.

However, we should take note that as reflected in Figure 7.8, roles assumed by game Web sites and offline events are different by game. This indicates game makers may manipulate channels of information gathering via marketing strategies.

The difference between offline event indicators above can be ascribed to the game maker's aggressive promotion activity. For example, Optic Communications (services Legend of Mir 3) executed marketing efforts employing offline channels.

> To facilitate expansion of Legend of Mir 3, we launched intensive promotion campaigns in Internet cafés located in cities where Legend

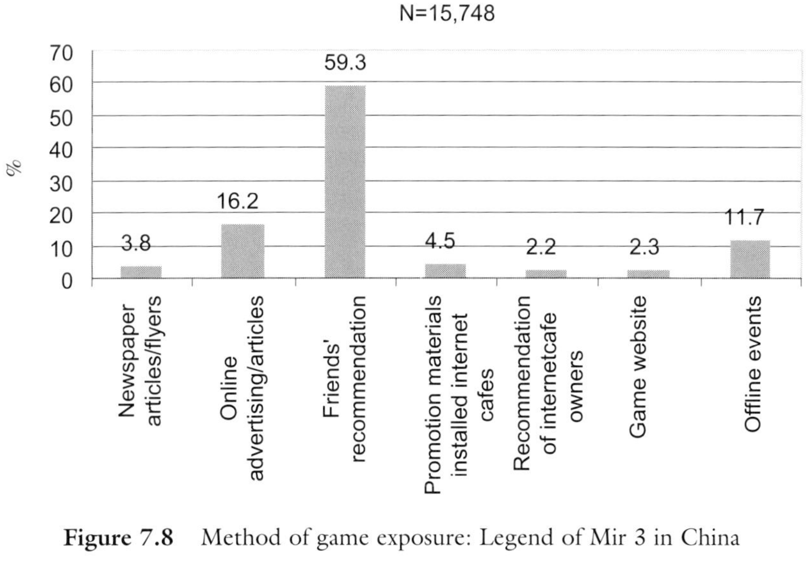

Figure 7.8 Method of game exposure: Legend of Mir 3 in China

> of Mir 2 users were concentrated. Huàzhōng, Héjiě, Shèngyǎn, and among them Shèngyǎn is near Sichuan and Chunking. And we held online events for new users. We drew new users with events offering rapid level advances or rewards when users within the server reach high levels.
>
> (Laoyao Ye, Product Manager of online game marketing at Optic Communications).

Another example is Oriental Interactive, which services A3, and their management of Internet cafés, offline event strongholds.

> We divided Chinese Internet cafés into three categories: mainstay Internet cafés, which offer nothing but A3; Internet cafés which only hang A3 signs; and ordinary Internet cafés. We intend to organize Internet cafés as a brand. We are preparing to launch A3 Inside, a comprehensive services brand. That way we could keep tabs on the sales network and change policies to steer profits to Internet cafés. We are trying to steer profits from cards being sold in Internet cafés via e-sales to the Internet cafés.
>
> (Chun Tong-hae, Marketing Director of Oriental Interactive).

3. PK preference

PK is an acronym for player killing and refers to stronger characters in online games attacking weaker characters. Users tend to personalize the character in the game hence when the character is subjected to attacks, praise or criticism, players feel as if they are subjected to those actions and/or sentiments in real life.

In the early stages of online games PKs were allowed unregulated, but now games have placed restrictions on PKs as stronger characters arbitrarily leveling an attack at weaker characters is looked upon askance. For example, when a PK occurs, sanctions are levied on the character that initiated the attack, and for a set duration their ID turns purple. Characters with purple IDs are referred to as "purple boys" in the game, and when a third party attacks purple boys, it does not qualify as a PK. If the character retaliates upon sustaining an attack, that attack is recognized as self-defense. For example 'Corum Online's PK system is described below.

PK Counter

— Character Disposition points: Ranges from –32767 points to +32767 points, and the higher the points the more benevolent the character is, the lower the more malicious.
— PK Counter (PC: PK Counter): Increases by 1 with execution of every PK; decreases by 1 if no crime is committed within a 36-hour period.
— RPCs (Repeated PK Counter): If PKs are continually executed within a 2-hour period increases by 1; if no crime is committed decreases by 1 every 160 hours.

PK Classification

— According to the kp awarded, character classified into standard level, new offender, repeated offender, felon.
— Upon character's resurrection following death, hp drops to 1 and sp reverts to the level at the time of death.
— Criminal-level characters are not eligible to invoke self-dense.

Conditions Immune to PKs (cases where self-defense is acknowledged)

a) When a character above the blue level sustains an attack from another party or another party's summoned object
b) When a character above the blue level sustains an attack from its own summoned object (monsters or guardians)
c) When a PK is executed while engaging in a PvP at an arena
d) PKs generated under dungeon/siege conditions
e) When a criminal-level character is killed
f) When a patron guardian is killed in a siege

PK Penalties

1) Upon character's death, the exp drop of criminals is steeper than that of ordinary characters.
2) Not allowed to enter villages in the future (plans to create a village to which only characters with a kao disposition are granted entry).

Source: <http://cafe.naver.com/corumnala.cafe>

According to the PK penalties description, characters executing PKs are deemed criminals, and banned from villages. This reflects real life wherein criminals cannot carry on a normal social life.

Despite such sanctions and moral stigma attached to them, PKs still exist in online games, and some users revel in them. PKs are also conceived differently according to region, and not deemed a criminal act but a duel between characters. Korean online games strictly ban PKs, but in the U.S. with games like WOW, and with Korean games being serviced in China, PKs are unregulated and left to the discretion of users.

Then let us observe PK frequency. Figure 7.9 is a study conducted on PK experience regardless of which role the respondent assumed.

Of MMORPG users in the U.S., 34.7% claimed to have player killing experience, and 65.3% had no PK experience, producing a ratio of 3:7. On the other hand, of MMORPG users in China, 91.3% claimed to

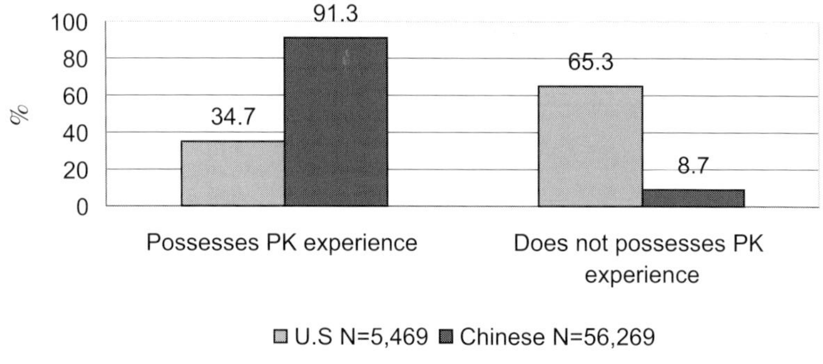

Figure 7.9 PK experience U.S. vs. China

have player killing experience, and 8.7% had no PK experience, producing a ratio of 9:1. The PK experience between users in China and the U.S. differed vastly. Let us examine which type of PKs the users execute.

Figure 7.10 examines the PK frequency between Korean and Chinese users. Results show a significant difference in PK frequency between the two user groups. Whereas 78.4% of Korean users almost never engage in PKs ("never" and "almost never"), a much lower figure of only 53.1% of Chinese users almost never arbitrarily kill other players.

Questions on PK degrees like the above is a direct reflection of users' PK conception. 39.6%, the highest proportion of Chinese users conceived PKs as "a defense mechanism for protecting characters." 24.9% of them responded "to flaunt the character's power," and 10.4% responded "a method other than hunts to acquire items." With only 25.1% of Chinese players replying "an abnormal behavior perpetrated by a minority," we can infer a negative conception of PKs is not prevalent.

Such results are vastly different from the results of a survey conducted on Korean Lineage 2 users. Lineage has one the most active PK occurrences among Korean MMORPGs.

However, 41.5% of Korean Lineage 2 users, almost twofold that of their Chinese counterparts, deemed PKs as abnormal behavior

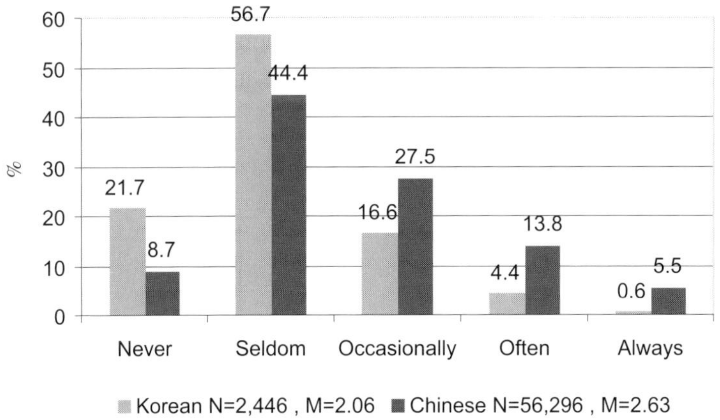

Figure 7.10 Intra-game PK degree comparison

* On a Scale of 1 to 5 (1 = Never, 5 = Always)

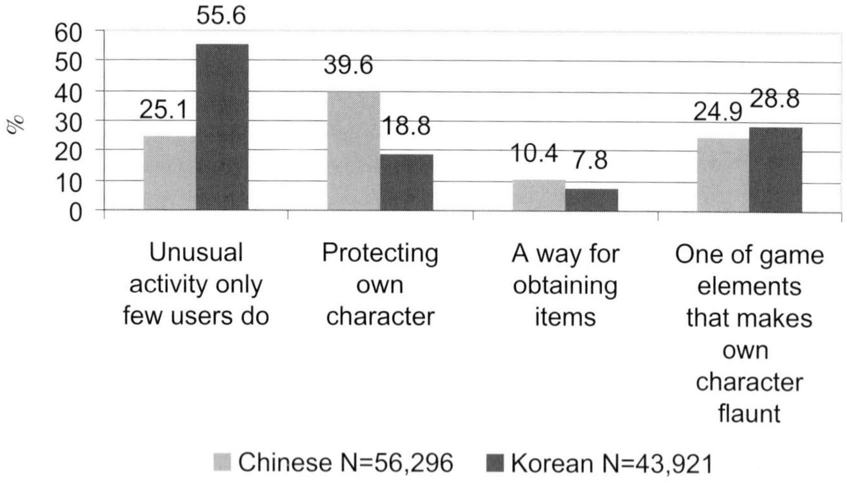

Figure 7.11 Users' PK conception

perpetrated by a minority. While Lineage users are among the most lenient Korean online game users as regards PKs, results indicate Chinese users are far more liberal in that regard.

Another amusing point is Chinese users claiming defense mechanism the primary reason for perpetrating PKs. This can be deemed as a passive PK, as in China, where PKs actively occur, the act of player killing as the product a synergistic relationship between the attacker and defender.

That is the difference between core users with extensive experience and light users with little experience. This is vividly reflected in a comparison between online game users in the U.S. and China. Users in the U.S. do not actively perpetrate PKs, but PKs are commonplace among Chinese users. The U.S. online game market is constructed centered on core users with extensive experience under their belts.

Such differences are intricately linked to whether users assign more value to the game process or the end result. Users in the U.S. or Japan tend to hold the process as a higher priority over the outcome. The process of completing a quest takes precedence for them over a rapid rise in character level. However, the outcome is a higher priority for Chinese users. Consequently, they desire to flaunt their achievements via character level, and that desire manifests as PKs.

As seen above, PKs surpass a simple duel between characters, and exhibit differences in culture and behavior patterns by user region and nationality.

4. Game styles

Items characters employ or are equipped with in online games serve two purposes. One is to facilitate character level ups, and the other is to flaunt to other users. With the former, when characters possess powerful weaponry or shields, their attack and defense ability rises in correlation and the objective is achieved more effectively.

With the latter case, characters can flaunt rare items to other players. In real life sports cars or luxury goods not only possess inherent value, but they are also purchased for the purpose of flaunting to others. These consumer goods serve an important function of vicariously flaunting the buyer's economic prowess and social status. Such is the mentality driving female office workers in their 20's who save for one year to buy the "it" purse.

Of the two motives, level up was a larger motive for Chinese users, and flaunting took a higher priority for Korean users. When considering level up help and flaunt in a parallel, 47.14%, the highest proportion of Chinese online game users responded "solely for level ups," and 6.46% responded "solely to flaunt to others." Fixation on item acquisition for level ups is an important distinguishing trait of Chinese users.

However, 35.5% of Korean users responded "solely for level ups," and 18.1% responded "solely to flaunt to others." When compared to users in China, the "solely for level ups" response by Korean users was significantly lower than the 47.14% of Chinese users, and the 18.1% "solely to flaunt to others" response was significantly higher than the 6.46% of Chinese users.

Based on these results, it can be inferred Korean users value inter-user relationship more than their Chinese counterparts.

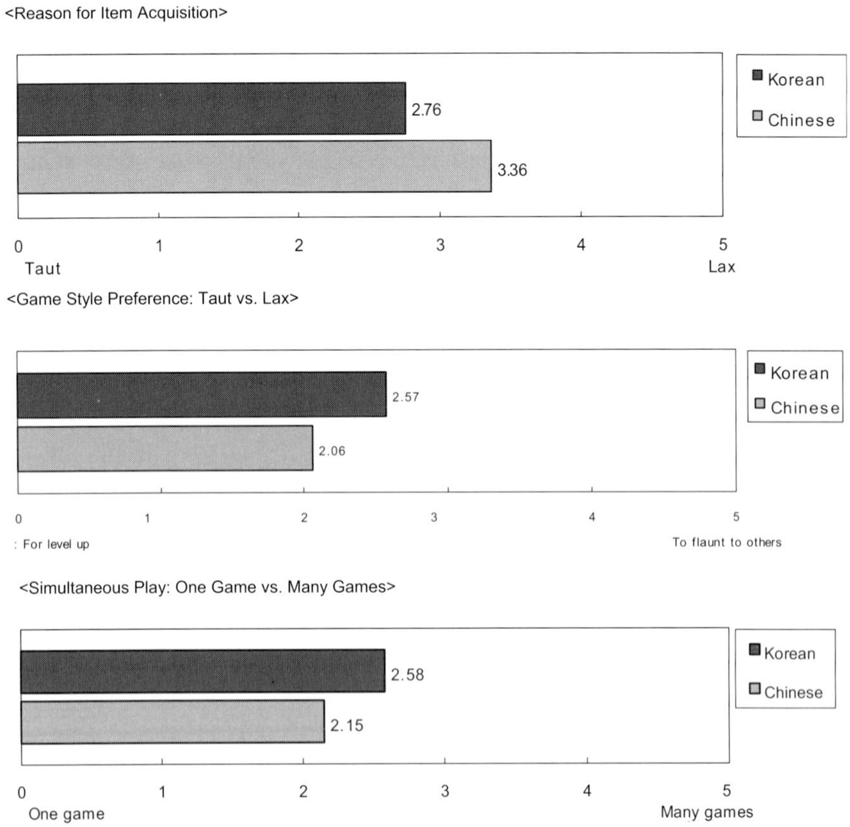

Figure 7.12 Reason for item acquisition: for level ups vs. to flaunt to others
Legend: China n = 56,356 Korea n = 1,283

Taut and focused vs. lax and low-key

Online game users in China are engaging in a more "lax gameplay" than their Korean counterparts. The median value of Chinese users is 3.36, significantly higher than the 2.76 of Koreans. This indicates the gameplay style of Chinese users is more lax involving frequent breaks.

Two factors contribute to such gameplay trends. One is the prevalence of macros. I have pointed out, in Chapter 6 in describing distinguishing attributes of the Chinese market, that macro use prevalent among users in China. When employing macros, even when playing

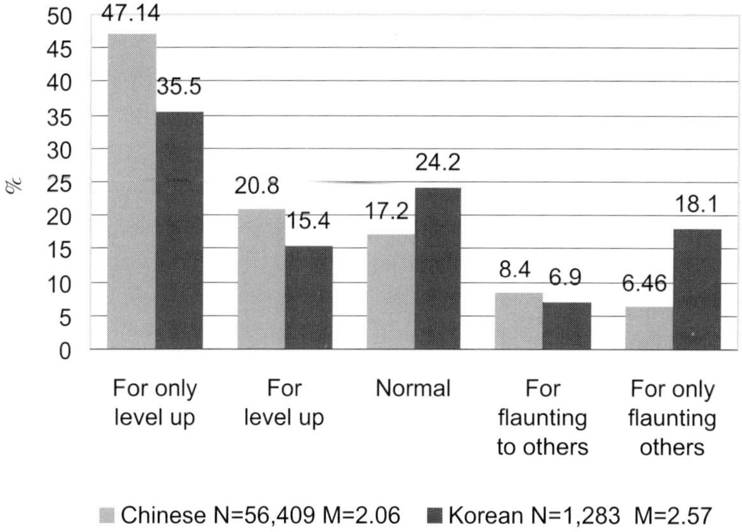

Figure 7.13 Chinese, Korean compare

for an extended duration, users need not grow tense or concentrate on the game. For example, the characters at hunting grounds need to remain wary at all times to ward off perpetual attacks, but if macros are employed, surrounding monsters are attacked automatically, and attacks are initiated at other passerby characters hence players can remain more at ease.

The other factor is the difference is gameplay duration. The average gameplay duration of Chinese players is 6 hours during the week, and 9 hours during weekends. In comparison, Korean Lineage players play on the average 4 hours on weekdays and 6 hours on weekends. Such difference in gameplay duration produces differences in degree of concentration and tension during gameplay. When playing for longer, periods physical requirements inevitably ease tension.

Concentrating on one game vs. simultaneously playing multiple games

Unlike video games, players cannot simultaneously play multiple online games. As an online game forms a virtual world, players can be excluded from the flow of the world when concentration is not afforded. If players have joined guilds, they can be expelled for not maintaining active gameplay. In other words, as individuals cannot hold down several jobs at once in real life, users of online games cannot spread themselves thin in multiple games.

In terms of such game concentration, Chinese users scored higher than their Korean counterparts. The median value of users in China was 2.15, and this indicates when Chinese users begin to play one game, they possess a strong tendency to play it continuously.

These results appear to be related to fees as well. In case of China a large number of users engage in gameplay at Internet cafés, and they are required to remit fees for gameplay and Internet café use at once. In other words, in a month, they pay roughly 30 yuan (about 25.5 USD) for monthly the gameplay fee and 3–4 yuan/hour for Internet café usage fee. Accordingly, playing multiple games at once increases the fee by as much. About 30%–40% of Chinese users are students. They have few opportunities of earning extra allowance by taking on part-time jobs hence paying for gameplay is a significant financial burden to them. The elements stated above contribute to the heightened concentration of Chinese users.

The fact that more free games are serviced in Korea than in China serves as another contributing factor. Game users in Korea set their usual games aside and shop around for free games.

5. Online game selection criteria

Figures 7.14 and 7.15 compare game selection criteria of MMORPG users in U.S. and China. 16.42% of U.S. MMORPG players responded "quest content," 15.57% "elaborate graphics (3D)," 9.72% "diverse and unique characters," and 4.8% "diverse items." U.S. MMORPG users hold quest content and design elements (elaborate graphics (3D), diverse and unique characters) as guidelines in game selection. Design-related elements such as elaborate graphics (3D), diverse and unique characters can be seen as resulting from console and PC-game influence. Chinese users display a contrast in this phenomenon.

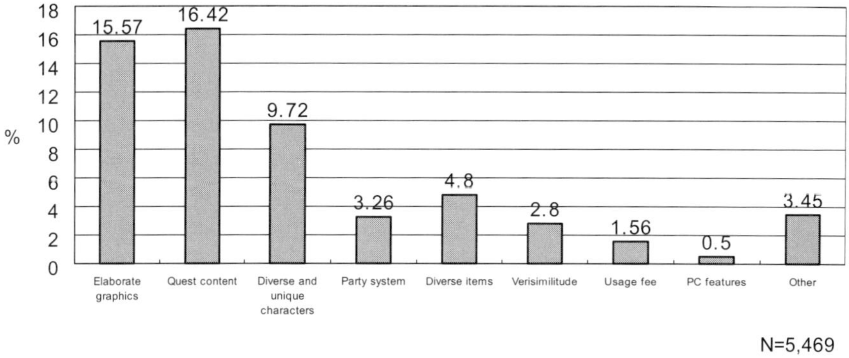

Figure 7.14 MMORPG selection criteria of U.S. users

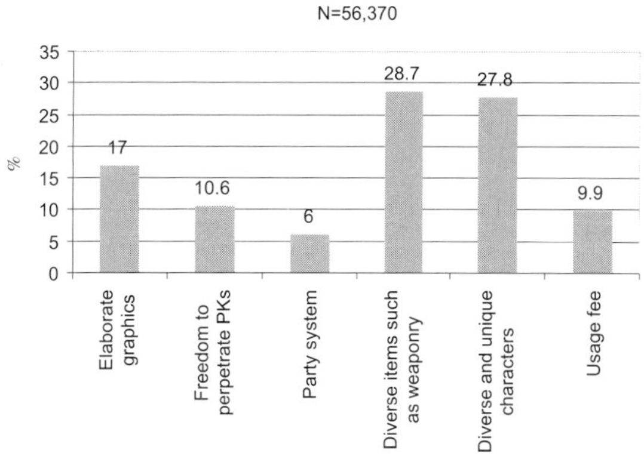

Figure 7.15 MMORPG selection criteria of Chinese users

Results indicate quest contents are an important factor in game selection, and this translates to Western gamers preferring games providing quests with diverse contents and pleasure generated by level up processes per se instead of "manual labor games" centered on simple repetition of monster hunts.

"Diverse items such as weaponry" was the most important criterion in game selection earning 28.7% of the responses of the MMORPG users in China surveyed, with "diverse and unique characters" coming in second at 27.8%, and "elaborate graphics" at 17%.

Accordingly, Chinese online game users consider items, characters and graphics as the most important factors in game selection.

Whereas quest content and elaborate graphics are a high priority for U.S. users, diverse items such as weaponry, and diverse and unique characters appears to take precedence for and Chinese users.

References

Books and papers

Abernathy, W J (1978). *The Productivity Dilemma: Roadback to Innovation in the Automobile Industry.* The Johns Hopkins University Press.

Abernathy, W J, K B Clark and A M Kantrow (1983). *Industrial Renaissance — Producing a Competitive Future for America.* New York, Basic Books Inc., Publishers.

Abernathy, W J and J M Utterback (1978). 'Patterns of industrial innovation.' *Technology Review*, 80(7), 40–47.

Castronova, E (2007). *Exodus to the Virtual World: How Online Fun Is Changing Reality.* St Martins Press.

Castronova, E (2006). *The Business and Culture of Online Games.* Chicago, University of Chicago Press.

Christensen, C M (1997). *The Innovator's Dilemma.* Harvard Business Press.

Christensen, C M and J Bower (1996). 'Customer power, strategic investment and the failure of leading firms.' *Strategic Management Journal*, 17, 197–218.

Clark, K B (1985). 'The interaction of design hierarchies and market concepts in technological evolution.' *Research Policy*, 14, 235–251.

David, P A (1997). 'Path dependence and the quest for historical economics: One more chorus of the ballad of QWERTY,' Discussion Papers in Economics and Social History, No. 20.

David, P A (1985). 'Clio and the economics of QWERTY.' *The American Economic Review*, 75(2), 332–336.

Henderson, R M (1992). 'Technological change and the management of architectural knowledge.' In *Transforming Education*, T A Kochan and M Useem (eds.), Oxford University Press.

Henderson, R M and K B Clark (1990). 'Architectural innovation: The reconfiguration of existing product technologies and the failure of established firms.' *Administrative Science Quarterly*, 35, 9–30.

Huizinger, J (1955). *Homo Ludens: A Study of the Play Element in Culture.* Beacon.

Jenkins, H (2002). 'Interactive audiences?' In *The New Media Book*, D Harries (ed.), BFI.

Lancaster, K (1991). *Modern Consumer Theory.* Edward Elgar.

Leonard-Barton, D (1992). 'Core capabilities and core rigidities: A paradox in managing new product development.' *Strategic Management Journal*, 13, 111–225.

Leonard-Barton, D (1988). 'Implementation as mutual adaptation of technology and organization.' *Research Policy*, 17, 251–267.

Marx, K (1992). *Capital 1.* Penguin Press.

Robert, T S (1971). *Innovative Behavior and Communication.* Holt, Rinehart and Winston.

Roh, S K and J H Wi (2007). *Development Process of Korean Online Game Industry.* Seoul National University Press. (in Korean)

Wi, J H (2008). *Industrial Development Strategy of Online Games.* Tsinghua University Press. (in Chinese)

Wi, J H (2007a). *Second Life Business Strategy.* Joong-Ang Books. (in Korean)

Wi, J H (2007b). *Innovation Strategy of Japanese Firms.* Jipmundang. (in Korean)

Wi, J H and N R Oh (2007). 'The influence of social self-efficacy on online game satisfaction.' *Journal of Korean Academy of Gaming*, 7(3), 69–78. (in Korean)

Wi, J H (2006a). *Business Strategy of Online Games.* Zeumidea. (in Korean)

Wi, J H (2006b). *The Research on Korean Online Game Business.* Toyokeizaishinposha. (in Japanese)

Wi, J H (2006c). 'Organization design for new product architecture development: Comparative analysis of Sharp and Casio on PDA development.' *Asian Journal of Technology Innovation.*

Wi, J H (2006d). 'Organizational design for the development of new product architecture: Comparative analysis of NEC and Toshiba.' *Journal of Technology Innovation.* (in Korean)

Wi, J H and E S Won (2006). 'Utilizing online game as an effective learning material — Consideration of a business strategy lecture by utilizing online game, 'Goonzu', for university students as a view of constructivism.' *Journal of Korean Academy of Gaming*, 6(4), 25–37. (in Korean)

Wi, J H (2003). 'The formation process of Korean online game industry.' In *Economic Analysis on Game Industry — Structure and Business Strategy on Game Software Industry*, J Shintaku *et al.* (eds.), Toyokeizaishinposha. (in Japanese)

Wi, J H and N R Oh (2006a). 'Influence of social support on virtual life satisfaction in online game.' *2006 Summer Congress of Korean Academic Society of Business Administration*, August 16–18, Phoenix Park, Pyeongchang, Korea. (in Korean)

Wi, J H and N R Oh (2006b). 'Social support and virtual life satisfaction in online games: Indigenous psychological analysis.' *2006 Annual Congress of Korean Psychological Association*, August 18–19, Seoul National University, Seoul, Korea. (in Korean)

Wi, J H (2004). 'Industrial formation process of Korean online game — Analysis on new industry formation through supplementary infrastructure.' *2002 Game Industry Journal.* (in Korean)

Wi, J H and M Nojima (2003). 'The analysis on difference of attributes of user group by path dependence — A comparison of 'Lineage' user in Korea and Japan.' *Akamon Management Review*, 1(1), 1–23. (in Japanese)

Wi, J H (2003). 'Beyond online games, beyond game-item trading.' *Akamon Management Review*, 2(1), 32–41. (in Japanese)

Newspapers, white papers and websites

CESA, *CESA Game Annual Report 2005.* (in Japanese)

CESA, *Forecast and Status of the Japanese Game Industry, 2007.* (in Japanese)

Contents Management Institute, *Strategic Approach to U.S. Market 2005.* (in Korean)

Electronic Times, Game as Education, Changing Preconceptions — Game Addiction, May 10, 2004. (in Korean)

Korean Game Development Agency, *The Rise of Korean Games 2007.* (in Korean)

Korean Game Development Agency, *Korea Game Annual Report 2002–2007.* (in Korean)

PC World, April 1995. (in Korean)

The Jurassic Park Normal — telnet 211.239.157.152

http://cafe.naver.com/corumnala.cafe

http://ccidconsulting.com
http://eq2.stationexchange.com/
http://fortress2.x2game.com/
http://global.goonzu.com/center/mall/product_status_list.asp?menu=86&status=N
http://kart.nexon.net/
http://knightonlineworld.gamersfirst.com/powerupstore.php#MagicBag
http://payment.roseonline.co.kr/
http://pcbang.ncsoft.co.kr/PayInfo/PayInfo.aspx
http://pcbang.nexon.co.kr/
http://www.gamebill.co.kr/
http://www.gamemeca.co.kr
http://www.gamemeca.co.kr
http://www.inven.co.kr
http://www.ItemBay.co.kr
http://www.joycity.com/mall/freestyle.asp
http://www.lineage2.com/
http://www.ncsoft.net/
http://www.onlinegameforum.org/
http://www.plaync.com
http://www.popsoft.com.cn
http://www.sayclub.com
http://www.thegamenews.com/
http://www.worldofwarcraft.co.kr/
http://www.itembay.co.kr

Index

Game title

Game company